Walter G. (Walter Gordon) Fox

Transition Curves

A Field Book for Engineers, Containing Rules and Tables for Laying Out....

Walter G. (Walter Gordon) Fox

Transition Curves
A Field Book for Engineers, Containing Rules and Tables for Laying Out....

ISBN/EAN: 9783337158743

Printed in Europe, USA, Canada, Australia, Japan

Cover: Foto ©berggeist007 / pixelio.de

More available books at **www.hansebooks.com**

TRANSITION CURVES.

A FIELD BOOK FOR ENGINEERS,

CONTAINING

RULES AND TABLES FOR LAYING OUT TRANSITION CURVES.

BY

WALTER G. FOX,

CIVIL ENGINEER.

NEW YORK:

D. VAN NOSTRAND COMPANY, PUBLISHERS,

23 MURRAY AND 27 WARREN STREETS.

1893.

PREFACE.

In this work the endeavor has been to condense the essential facts and principles constituting the theory of the transition curve and to demonstrate how the curve can be conveniently laid out in field practice.

The same method has been adopted as is usually employed in running circular curves so that the field operations may be more readily comprehended and performed.

WALTER G. FOX.

NEW YORK, October 3d, 1893.

CHAPTER I

INTRODUCTION.

Near the ends of a railway curve the curvature should be gradually diminished to enable trains at high speed to deviate gradually from the tangent and avoid any snock caused by changing too suddenly from a straight line to a sharp curve.

A transition curve is generally used to connect the circular curve with the tangent as it makes easy transition between them.

Moreover, as the outer rail of a curved track is elevated above the inner rail, a transition curve will unite better with the tangent; for the super elevation of 'the outer rail is proportional to the degree of curve which is least at the initial point.

CHAPTER II.

A transition curve is a compound curve with many changes of radius. It has nearly the form of the cubic parabola.

The curve is laid out on chords with the transit in the same manner as circular curves. A uniform chord length of 10 feet has been arbitrarily assumed to facilitate calculation, and the degree of curve changed at the end of every chord. The deflection angles have been computed and will be found in the tables.

To compute the deflection angles it was necessary to have the co-ordinates of every point of compound curve. The sines and cosines of the different angles between the tangent and the chords of the curves multiplied by 10 will give the required co-ordinates.

In Fig. I. T is the beginning of the curve. The distance measured along the tangent in the direction of L is the latitude and in the direction of D at a right angle

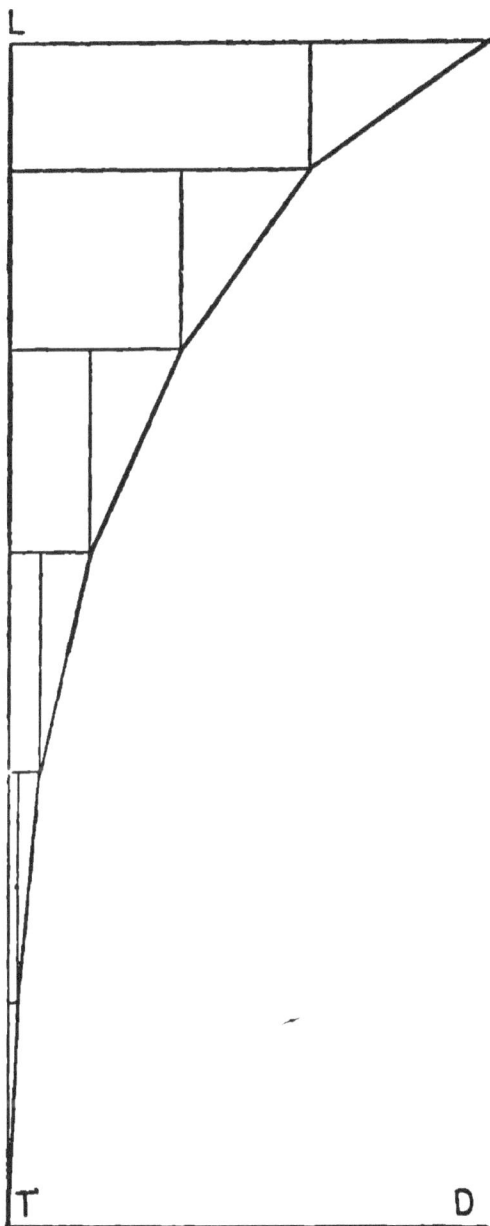

Fig. 1.

with the tangent is the departure of the various points of compound curves.

For example let us refer to Table I., which begins with a 0° 10′ curve and compounds into a sharper curve at the end of each chord, the degree of curve changing to 0° 20′, 0° 30′, 0° 40′, etc. See Fig. 2.

Dist	Degree of curve.	Angle of chord.	Cosine x 10.	Sine x 10.	Co-ordinates	
					Latitude.	Departure.
10	0° 10′	0° 0′ 30″	10.000000	0.001454	10.000000	0.001454
20	0° 20′	0° 2′ 00″	9.999998	0.005818	19.999998	0.007272
30	0° 30′	0° 4′ 30″	9.999991	0.013090	29.999989	0.020362
40	0° 40′	0° 8′ 00″	9.999973	0.023271	39.999962	0.043633
50	0° 50′	0° 12′ 30″	9.999933	0.036361	49.999895	0.079994

Having calculated the co-ordinates of the different points the deflection angles were determined by

$$\frac{D}{L} = \tan. \ T.$$

Example: To find the deflection angle for the first point in Table I.

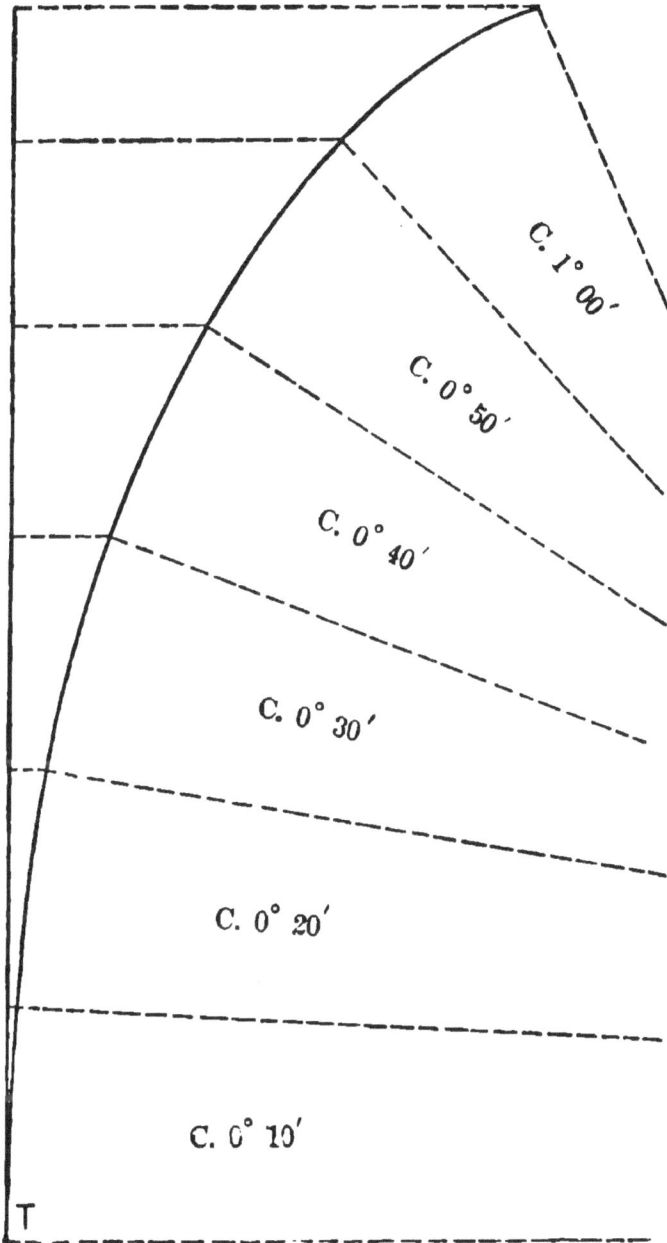

C. 1° 00'

C. 0° 50'

C. 0° 40'

C. 0° 30'

C. 0° 20'

C. 0° 10'

T

Fig. 2.

| Dep., | 0.001454 | log. | — 3.1625644 |
| Lat., | 10.000000 | log. | 1.0000000 |

Ans. . Def. angle, 0° 0′ 30″ log. tan. 6.1625644

The difference in the degree of curve between the different arcs of the transition curve should be maintained as nearly as possible between the terminating arc and the circular curve.

In the field the transit should be placed at T and the angles deflected from the tangent. In case there should be an obstacle making a point invisible at T, the next point should be measured on the long chord. The intermediate point can then be easily set.

To find the length of the long chord to any point on a curve, Fig. 3.

Let C be the length of chord, L the latitude of the point, and A the deflection angle.

$$C = \frac{L}{\cos A}$$

Example:

To find the length of the long chord to the 8th point of the curve in Table III.

11

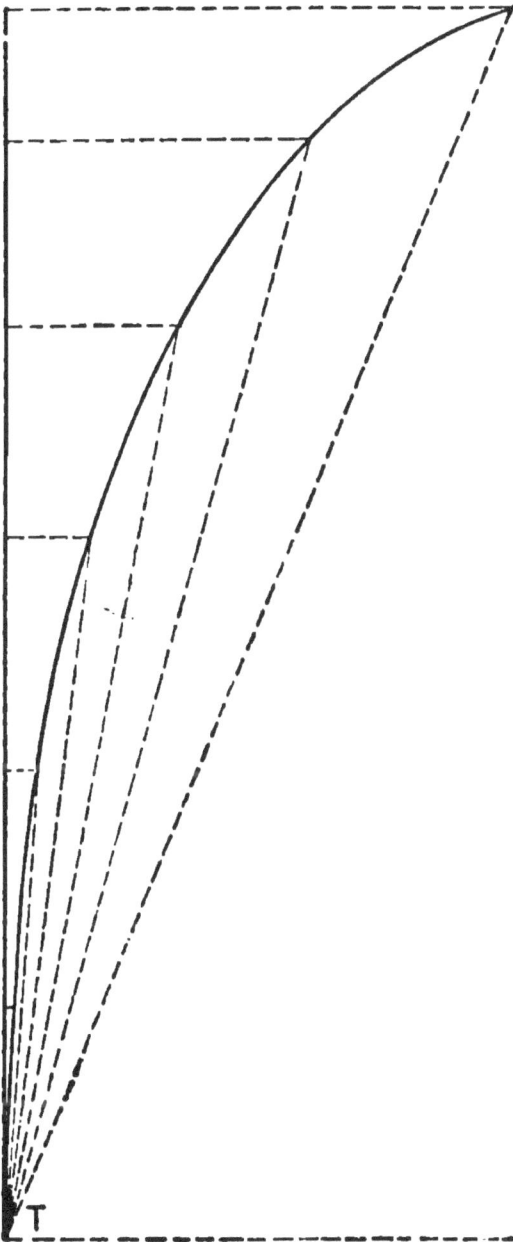

Fig. 3.

	Lat.	79.991	log.	1.9030411
	Def. angle,	0° 38′ 15″	log. cos.	9.9999731
Ans.	Long chord,	79.996	log.	1.9030680

CHAPTER III.

The curvature increases more rapidly in each succeeding table.

Table I. begins with a 0° 10′ curve which is the difference in curvature between the first arc of each table, the degrees of curve being 0° 10′, 0° 20′, 0° 30′, etc. The degree of curve of the first arc is equal to the difference between all the other arcs in the same table, so that the sixth arc in Table I. is a 1° curve and the sixth arc in Table VI. is a 6° curve.

Column I. gives the length of curve according to the central angle consumed which will be found on the same line in a different column, and also the other corresponding parts of the tables under their respective headings.

Column II. has the degree of curve of each separate arc.

Column III. contains the deflection angles. It will be observed that the value of the angles has been calculated to the nearest second. Such accuracy is not expected in practice but was necessary in order to correctly determine the long chords.

Column IV. gives the central angles of the curves.

Columns V. and VI. have the co-ordinates of every point of compound curve, By using the co-ordinates in place of the deflection angles the curve may be laid out with offsets from the tangent.

Column VII. has the long chords to every point on the curves.

Column VIII. contains numbers by which the tangents of half the intersection angles of the curves should be multiplied and the product added to the tangent distances as a correction.

CHAPTER IV.

PROBLEMS.

Given, a 5 degree curve, to run a transition curve connecting it with a tangent having an intersection angle of 20 degrees.

If we select the curve from Table VI. its length will be 40 feet, consisting of 4 arcs, and the central angle will be 1 degree. As this must be repeated at the other end of the curve, the central angle should be multiplied by 2 and the sum subtracted from the intersection angle. The remainder will be the central angle of the circular curve. To find its length divide the central angle by the degree of curve.

Example:

Intersection angle,	20°
Central angle of trans. curve $\times 2$,	2°
	5/ 18°
Length of circular curve,	360 ft.

To find the tangent distance V.

Let M. be the length of circular curve
that the central angle of transition curve
consumes, a d the apex distance of circular curve and L the latitude of the terminal point of the transition curve.

$$V = ad + L - M.$$

Example:

Apex. dist. of 5° curve for 20°,	202.120
Lat. of ter'l point of trans. curve,	39.998
	242.118
Length of 5° curve to consume 1°,	20.000
Ans. V = Tan. distance,	222.118

A correction should be added to the product which will not materially change the
result unless the length of curve exceeds
70 feet, or the intersection angle contains
more than 50 degrees.

Rule: Multiply the tangent of half the
intersection angle by the distance in column
8 corresponding to the terminal point of
the curve.

Example:

1/2 intersection angle = 10°
Tan. = .1763 × .086 =   -   .015
then V =   -   -   -   - 222.118
Correction   -   -   -   .015
Tan. dist.   -   -   - 222.133

With curves of short radius there is a
slight error in the preceding formula, as M
does not always have exactly its true value.
The error, however, is too small to take
into account in field practice. In the above
example it is only one-thousandth of a
foot. In a ten degree curve the error is
about three-tenths.

The true value of M can be determined
by multiplying the cosine of half the central angle of transition curve by the length
of circular curve it has consumed.

Example:

1/2 central angle of trans.
curve = 0° 30′ log. cos. 9.9999835
Length of circular curve consumed = 20 feet   log. 1.3010300

Ans. M. = 19.999 ft.   log. 1.3010135

Having measured the tangent distances,
the transition curves should be first laid
out and then connected by running the circular curve. See Fig. 4.

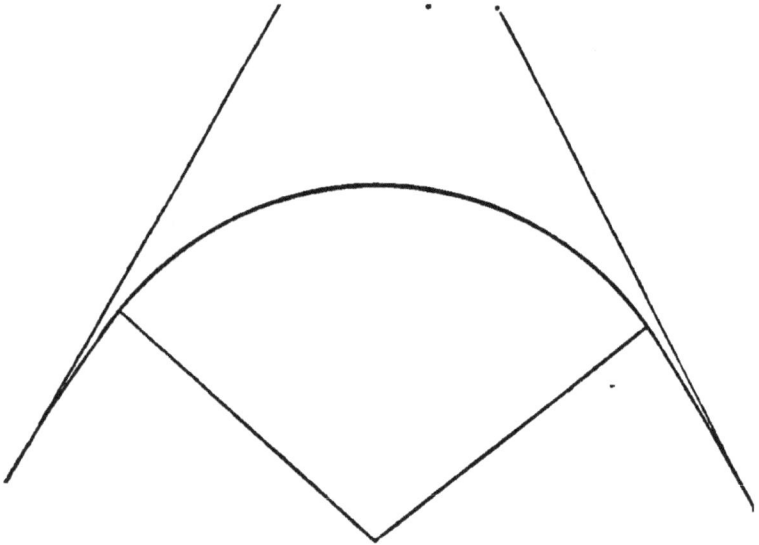

Fig. 4.

If there are two circular curves of different radii, their ends will be a short distance apart, the sharper curve being on the
inner side, which can then be eased off to
unite with the other curve. See Fig. 5.

On railway location it may be desirable
to first lay out the circular curve, neglect-
ing the transition curves until the time of
construction.

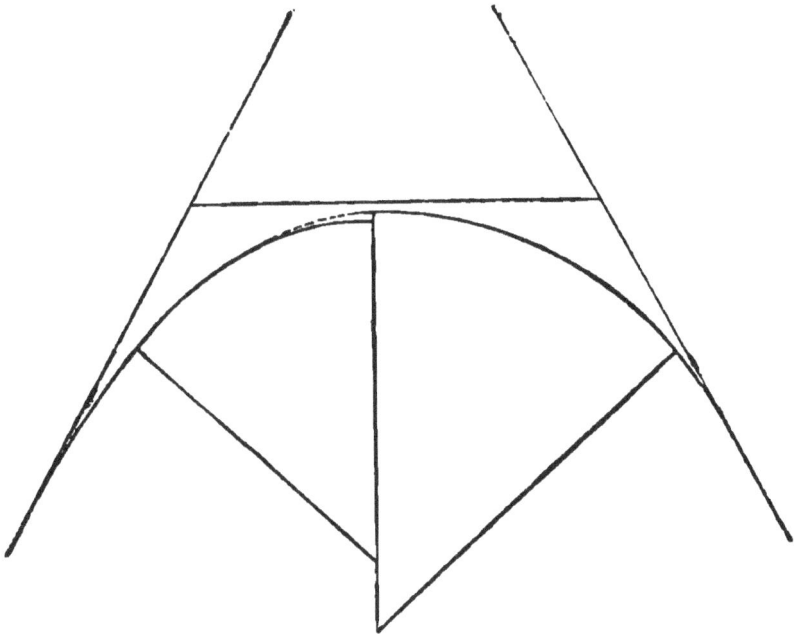

Fig. 5.

In this case the two terminal points of
the transition curves should be located with
their co-ordinates, the transit set over
either point and a foresight taken on the
other and the curve run in.

Compound curves should not be located
in this manner, as a backsight is then required.

TRANSITION CURVE TABLES.

TABLE I.

Length of Curve.	Degree of Curve.	Deflection Angle.	Central Angle.	Co-ordinates Latitude.	Co-ordinates Departure.	Long Chord.	Corr'n for Tang.Dist.
10	0° 10′	0° 0′ 30″	0° 1′ 0″	10.000	0.001	10.000	.000
20	0° 20′	0° 1′ 12″	0° 3′ 0″	20.000	0.007	20.000	.004
30	0° 30′	0° 2′ 17″	0° 6′ 0″	30.000	0.020	30.000	.007
40	0° 40′	0° 3′ 41″	0° 10′ 0″	40.000	0.043	40.000	.014
50	0° 50′	0° 5′ 30″	0° 15′ 0″	50.000	0.080	50.000	.029
60	1° 00′	0° 7′ 33″	0° 21′ 0″	60.000	0.132	60.000	.045
70	1° 10′	0° 9′ 58″	0° 28′ 0″	69.999	0.203	69.999	.061
80	1° 20′	0° 12′ 43″	0° 36′ 0″	79.999	0.296	79.999	.086
90	1° 30′	0° 15′ 49″	0° 45′ 0″	89.998	0.414	89.999	.126
100	1° 40′	0° 19′ 15″	0° 55′ 0″	99.997	0.560	99.998	.168
110	1° 50′	0° 23′ 0″	1° 6′ 0″	109.995	0.736	109.997	.208
120	2° 00′	0° 27′ 4″	1° 18′ 0″	119.993	0.945	119.996	.265
130	2° 10′	0° 31′ 30″	1° 31′ 0″	129.990	1.191	129.995	.341
140	2° 20′	0° 36′ 15″	1° 45′ 0″	139.986	1.476	139.994	.417
150	2° 30′	0° 41′ 19″	2° 0′ 0″	149.981	1.803	149.992	.494
160	2° 40′	0° 46′ 44″	2° 16′ 0″	159.974	2.175	159.989	.593
170	2° 50′	0° 52′ 30″	2° 33′ 0″	169.965	2.596	169.985	.717
180	3° 00′	0° 58′ 35″	2° 51′ 0″	179.954	3.067	179.980	.843
190	3° 10′	1° 4′ 59″	3° 10′ 0″	189.940	3.591	189.974	.967
200	3° 20′	1° 11′ 45″	3° 30′ 0″	199.923	4.173	199.966	1.119
210	3° 30′	1° 18′ 50″	3° 51′ 0″	209.903	4.814	209.958	1.302
220	3° 40′	1° 26′ 14″	4° 13′ 0″	219.878	5.517	219.947	1.487
230	3° 50′	1° 34′ 0″	4° 36′ 0″	229.848	6.286	229.934	1.671
240	4° 00′	1° 42′ 5″	5° 0′ 0″	239.814	7.123	239.920	1.889

TABLE II.

Length of Curve.	Degree of Curve.	Deflection Angle.	Central Angle.	Co-ordinates Lati-tude.	Co-ordinates Depar-ture.	Long Chord.	Corr'n for Tang.Dist.
10	0° 20'	0° 1' 0''	0° 2' 0''	10.000	0.003	10.000	.002
20	0° 40'	0° 2' 24''	0° 6' 0''	20.000	0.014	20.000	.005
30	1° 00'	0° 4' 35''	0°12' 0''	30.000	0.040	30.000	.014
40	1° 20'	0° 7' 28''	0°20' 0''	40.000	0.087	40.000	.029
50	1° 40'	0° 11' 0''	0°30' 0''	50.000	0.160	50.000	.051
60	2° 00'	0° 15' 7''	0°42' 0''	59.999	0.264	59.999	.081
70	2° 20'	0° 19' 59''	0°56' 0''	69.998	0.407	69.999	.122
80	2° 40'	0° 25' 29''	1°12' 0''	79.996	0.593	79.998	.174
90	3° 00'	0° 31' 40''	1°30' 0''	89.993	0.829	89.996	.240
100	3° 20'	0° 38' 30''	1°50' 0''	99.989	1.120	99.995	.320
110	3° 40'	0° 45' 58''	2°12' 0''	109.983	1.471	109.993	.415
120	4° 00'	0° 54' 9''	2°36' 0''	119.974	1.890	119.989	.529
130	4° 20'	1° 3' 0''	3°02' 0''	129.962	2.382	129.983	.662
140	4° 40'	1° 12' 29''	3°30' 0''	139.946	2.951	139.977	.813
150	5° 00'	1° 22' 39''	4° 0' 0''	149.924	3.605	149.967	.988
160	5° 20'	1° 33' 29''	4°32' 0''	159.897	4.349	159.956	1.185
170	5° 40'	1° 44' 59''	5° 6' 0''	169.861	5.189	169.940	1.407
180	6° 00'	1° 57' 9''	5°42' 0''	179.817	6.130	179.921	1.655
190	6° 20'	2° 9' 58''	6°20' 0''	189.762	7.178	189.897	1.930
200	6° 40'	2° 23' 28''	7° 0' 0''	199.694	8.339	199.868	2.234
210	7° 00'	2° 37 39''	7°42' 0''	209.612	9.619	209.832	2.569
220	7° 20'	2° 52' 28''	8°26' 0''	219.513	11.022	219.789	2.934
230	7° 40'	3° 7' 58''	9°12' 0''	229.395	12.555	229.738	3.332
240	8° 00'	3° 24' 6''	10° 0' 0''	239.255	14.222	239.677	3.763

TABLE III.

Length of Curve.	Degree of Curve.	Deflection Angle.	Central Angle.	Co-ordinates Latitude.	Departure.	Long Chord.	Corr'n for Tang.Dist.
10	0° 30'	0° 1' 30''	0° 3' 0''	10.000	0.004	10.000	.003
20	1° 00'	0° 3' 47''	0° 9' 0''	20.000	0.022	20.000	.010
30	1° 30'	0° 6' 59''	0° 18' 0''	30.000	0.061	30.000	.022
40	2° 00'	0° 11' 15''	0° 30' 0''	40.000	0.131	40.000	.044
50	2° 30'	0° 16' 30''	0° 45' 0''	49.999	0.240	49.999	.080
60	3° 00'	0° 22' 45''	1° 3' 0''	59.998	0.397	59.999	.126
70	3° 30'	0° 30' 0''	1° 24' 0''	69.995	0.611	69.997	.183
80	4° 00'	0° 38' 15''	1° 48' 0''	79.991	0.890	79.996	.262
90	4° 30'	0° 47' 29''	2° 15' 0''	89.985	1.243	89.993	.366
100	5° 00'	0° 57' 44''	2° 45' 0''	99.975	1.679	99.989	.487
110	5° 30'	1° 8' 59''	3° 18' 0''	109.962	2.207	109.984	.624
120	6° 00'	1° 21' 14''	3° 54' 0''	119.942	2.835	119.975	.793
130	6° 30'	1° 34' 30''	4° 33' 0''	129.915	3.572	129.964	1.001
140	7° 00'	1° 48' 44''	5° 15' 0''	139.878	4.426	139.948	1.231
150	7° 30'	2° 3' 59''	6° 0' 0''	149.830	5.406	149.927	1.480
160	8° 00'	2° 20' 14''	6° 48' 0''	159.768	6.521	159.901	1.776
170	8° 30'	2° 37' 28''	7° 39' 0''	169.688	7.778	169.866	2.130
180	9° 00'	2° 55' 42''	8° 33' 0''	179.589	9.187	179.824	2.491
190	9° 30'	3° 14' 57''	9° 30' 0''	189.465	10.756	189.770	2.889
200	10° 00'	3° 35' 11''	10° 30' 0''	199.313	12.492	199.704	3.342
210	10° 30'	3° 56' 25''	11° 33' 0''	209.128	14.405	209.623	3.855
220	11° 00'	4° 18' 39''	12° 39' 0''	218.906	16.501	219.527	4.399
230	11° 30'	4° 41' 52''	13° 48' 0''	228.641	18.789	229.411	4.973
240	12° 00'	5° 6' 5''	15° 0' 0''	238.327	21.276	239.275	5.612

TABLE IV.

Length of Curve.	Degree of Curve.	Deflection Angle.	Central Angle.	Co-ordinates Latitude.	Depar ture.	Long Chord.	Corr'n for Tang.Dist.
10	0° 40′	0° 2′ 0″	0° 4′ 0″	10.000	0.006	10.000	.003
20	1° 20′	0° 4′ 59″	0° 12′ 0″	20.000	0.029	20.000	.012
30	2° 00′	0° 9′ 17″	0° 24′ 0″	30.000	0.081	30.000	.029
40	2° 40′	0° 14′ 57″	0° 40′ 0″	39.999	0.174	39.999	.058
50	3° 20′	0° 22′ 0″	1° 0′ 0″	49.998	0.320	49.999	.102
60	4° 00′	0° 30′ 19″	1° 24′ 0″	59.996	0.529	59.998	.162
70	4° 40′	0° 39′ 59″	1° 52′ 0″	69.992	0.814	69.996	.244
80	5° 20′	0° 50′ 58″	2° 24′ 0″	79.985	1.186	79.993	.348
90	6° 00′	1° 3′ 18″	3° 0′ 0″	89.974	1.657	89.989	.479
100	6° 40′	1° 16′ 59″	3° 40′ 0″	99.957	2.239	99.982	.639
110	7° 20′	1° 31′ 58″	4° 24′ 0″	109.932	2.942	109.971	.831
120	8° 00′	1° 48′ 19″	5° 12′ 0″	119.897	3.779	119.956	1.057
130	8° 40′	2° 5′ 59″	6° 04′ 0″	129.849	4.761	129.936	1.321
140	9° 20′	2° 24′ 58″	7° 0′ 0″	139.784	5.898	139.908	1.625
150	10° 00′	2° 45′ 18″	8° 0′ 0″	149.698	7.204	149.871	1.972
160	10° 40′	3° 6′ 58″	9° 4′ 0″	159.588	8.688	159.824	2.365
170	11° 20′	3° 29′ 57″	10° 12′ 0″	169.446	10.361	169.762	2.805
180	12° 00′	3° 54′ 15″	11° 24′ 0″	179.269	12.235	179.686	3.296
190	12° 40′	4° 19′ 54″	12° 40′ 0″	189.050	14.320	189.591	3.841
200	13° 20′	4° 46′ 52″	14° 0′ 0″	198.780	16.626	199.474	4.439
210	14° 00′	5° 15′ 9″	15° 24′ 0″	208.453	19.163	209.332	5.094
220	14° 40′	5° 44′ 45″	16° 52′ 0″	218.059	21.942	219.160	5.809
230	15° 20′	6° 15′ 41″	18° 24′ 0″	227.589	24.971	228.955	6.585
240	16° 00′	6° 47′ 56″	20° 0′ 0″	237.033	28.260	238.711	7.422

TABLE V.

Length of Curve.	Degree of Curve.	Deflection Angle.	Central Angle.	Co-ordinates		Long Chord.	Corr'n for Taug.Disc.
				Latitude.	Depar ture.		
10	0° 50'	0° 2' 30''	0° 5' 0''	10.000	0.007	10.000	.004
20	1° 40'	0° 6' 11''	0° 15' 0''	20.000	0.036	20.000	.016
30	2° 30'	0° 11' 41''	0° 30' 0''	29.999	0.102	29.999	.037
40	3° 20'	0° 18' 44''	0° 50' 0''	39.999	0.218	39.999	.073
50	4° 10'	0° 27' 30''	1° 15' 0''	49.997	0.400	49.998	.131
60	5° 00'	0° 37' 52''	1° 45' 0''	59.994	0.661	59.997	.207
70	5° 50'	0° 50' 0''	2° 20' 0''	69.987	1.018	69.994	.305
80	6° 40'	1° 3' 44''	3° 0' 0''	79.977	1.483	79.990	.436
90	7° 30'	1° 19' 10''	3° 45' 0''	89.959	2.072	89.982	.606
100	8° 20'	1° 36' 14''	4° 35' 0''	99.933	2.798	99.972	.806
110	9° 10'	1° 54' 59''	5° 30' 0''	109.894	3.677	109.955	1.038
120	10° 00'	2° 15' 23''	6° 30' 0''	119.839	4.722	119.932	1.321
130	10° 50'	2° 37' 28''	7° 35' 0''	129.764	5.948	129.900	1.659
140	11° 40'	3° 1' 13''	8° 45' 0''	139.662	7.369	139.856	2.039
150	12° 30'	3° 26' 37''	10° 0' 0''	149.529	8.998	149.799	2.461
160	13° 20'	3° 53' 41''	11° 20' 0''	159.356	10.849	159.725	2.950
170	14° 10'	4° 22' 24''	12° 45' 0''	169.136	12.935	169.630	3.509
180	15° 00'	4° 52' 46''	14° 15' 0''	178.860	15.269	179.510	4.119
190	15° 50'	5° 24' 48''	15° 50' 0''	188.517	17.865	189.361	4.781
200	16° 40'	5° 58' 29''	17° 30' 0''	198.097	20.733	199.179	5.521
210	17° 30'	6° 33' 49''	19° 15' 0''	207.587	23.885	208.956	6.344
220	18° 20'	7° 10' 47''	21° 5' 0''	216.974	27.332	218.689	7.223
230	19° 10'	7° 49' 23''	23° 0' 0''	226.243	31.085	228.368	8.157
240	20° 00'	8° 29' 39''	25° 0' 0''	235.379	35.153	237.989	9.180

TABLE VI.

Length of Curve.	Degree of Curve.	Deflection Angle.	Central Angle.	Co-ordinates Latitude.	Co-ordinates Departure.	Long Chord.	Corr'n for Taug. Dist.
10	1° 00′	0° 3′ 0″	0° 6′ 0″	10.000	0.008	10.000	.004
20	2° 00′	0° 7′ 23″	0° 18′ 0″	20.000	0.043	20.000	.017
30	3° 00′	0° 13′ 58″	0° 36′ 0″	29.999	0.122	29.999	.043
40	4° 00′	0° 22′ 26″	1° 0′ 0″	39.998	0.261	39.999	.086
50	5° 00′	0° 33′ 0″	1° 30′ 0″	49.996	0.480	49.998	.153
60	6° 00′	0° 45′ 30″	2° 6′ 0″	59.991	0.794	59.996	.244
70	7° 00′	0° 59′ 58″	2° 48′ 0″	69.982	1.221	69.992	.366
80	8° 00′	1° 16′ 28″	3° 36′ 0″	79.966	1.779	79.985	.523
90	9° 00′	1° 35′ 0″	4° 30′ 0″	89.941	2.486	89.975	.719
100	10° 00′	1° 55′ 28″	5° 30′ 0″	99.903	3.357	99.959	.958
110	11° 00′	2° 17′ 58″	6° 36′ 0″	109.848	4.411	109.936	1.245
120	12° 00′	2° 42′ 29″	7° 48′ 0″	119.769	5.665	119.903	1.584
130	13° 00′	3° 8′ 57″	9° 06′ 0″	129.660	7.134	129.856	1.978
140	14° 00′	3° 37′ 26″	10° 30′ 0″	139.514	8.836	139.793	2.431
150	15° 00′	4° 7′ 55″	12° 00′ 0″	149.322	10.787	149.711	2.947
160	16° 00′	4° 40′ 23″	13° 36′ 0″	159.073	13.003	159.603	3.531
170	17° 00′	5° 14′ 49″	15° 18′ 0″	168.757	15.498	169.467	4.183
180	18° 00′	5° 51′ 15″	17° 06′ 0″	178.360	18.288	179.295	4.908
190	19° 00′	6° 29′ 39″	19° 0′ 0″	187.868	21.386	189.081	5.706
200	20° 00′	7° 10′ 2″	21° 0′ 0″	197.265	24.806	198.818	6.582
210	21° 00′	7° 52′ 24″	23° 6′ 0″	206.534	28.561	208.499	7.538
220	22° 00′	8° 36′ 42″	25° 18′ 0″	215.655	32.660	218.114	8.571
230	23° 00′	9° 22′ 58″	27° 36′ 0″	224.608	37.114	227.654	9.683
240	24° 00′	10° 11′ 10″	30° 0′ 0″	233.371	41.932	237.108	10.874

TABLE OF TANGENT DISTANCES

FOR

CIRCULAR RAILROAD CURVES.

This table contains the tangent distances to a 1 degree curve for every two minutes of intersection angle to 90 degrees.

The tangent distance to any other degree of curve may be determined by dividing the number corresponding to the intersection angle by the degree of curve.

′	0°	1°	2°	3°	′
0	0.00	50.00	100.00	150.07	0
2	1.67	51.67	101.67	151.74	2
4	3.33	53.33	103 34	153.41	4
6	5.00	55.00	105.01	155.08	6
8	6.67	56.67	106.68	156.75	8
10	8.33	58.33	108.35	158.42	10
12	10.00	60.00	110.02	160.09	12
14	11.67	61.67	111.69	161.76	14
16	13.33	63.33	113.36	163.43	16
18	15.00	65.00	115.02	165.09	18
20	16.67	66.67	116.69	166.76	20
22	18.33	68.33	118.36	168.43	22
24	20.00	70.00	120.03	170.10	24
26	21.67	71.67	121.70	171.77	26
28	23.33	73.33	123.37	173.44	28
30	25.00	75.00	125.03	175.10	30
32	26.67	76.67	126.70	176.72	32
34	28.33	78.33	128.37	178.39	34
36	30.00	80.00	130.04	180.06	36
38	31.67	81.67	131.71	181.73	38
40	33.33	83.33	133.38	183.40	40
42	35.00	85.00	135.05	185.07	42
44	36.67	86.67	136.72	186.74	44
46	38.33	88.33	138.38	188.40	46
48	40.00	90.00	140.05	190.07	48
50	41.67	91.67	141.72	191.74	50
52	43.33	93.33	143.39	193.41	52
54	45.00	95.00	145.06	195.08	54
56	46.67	96.67	146.73	196.75	56
58	48.33	98.33	148.40	198.42	58
60	50.00	100.00	150.07	200.09	60

′	4°	5°	6°	7°	′
0	200.09	250.17	300.30	350.44	0
2	201.76	251.84	301.97	352.11	2
4	203.43	253.51	303.64	353.79	4
6	205.10	255.18	305.31	355.46	6
8	206.77	256.85	306.98	357.13	8
10	208.44	258.52	308.65	358.81	10
12	210.11	260.20	310.32	360.48	12
14	211.77	261.86	311.99	362.15	14
16	213.45	263.54	313.66	363.83	16
18	215.11	265.20	315.33	365.50	18
20	216.78	266.87	317.00	367.17	20
22	218.45	268.54	318.67	368.85	22
24	220.12	270.21	320.34	370.52	24
26	221.79	271.88	322.01	372.19	26
28	223.46	273.54	323.68	373.86	28
30	225.13	275.21	325.35	375.54	30
32	226.80	276.88	327.02	377.22	32
34	228.47	278.55	328.69	378.89	34
36	230.14	280.23	330.37	380.57	36
38	231.81	281.90	332.04	382.24	38
40	233.48	283.57	333.71	383.92	40
42	235.15	285.24	335.38	385.60	42
44	236.82	286.91	337.05	387.27	44
46	238.48	288.59	338.73	388.95	46
48	240.15	290.26	340.40	390.62	48
50	241.82	291.93	342.07	392.30	50
52	243.49	293.60	343.74	393.98	52
54	245.16	295.27	345.41	395.65	54
56	246.83	296.95	347.08	397.33	56
58	248.50	298.62	348.76	399.01	58
60	250.17	300.30	350.44	400.70	60

′	8°	9°	10°	11°	′
0	400.70	450.95	501.32	551.74	0
2	402.37	452.63	503.00	553.42	2
4	404.05	454.31	504.68	555.10	4
6	405.72	455.98	506.36	556.78	6
8	407.39	457.66	508.04	558.46	8
10	409.06	459.34	509.72	560.14	10
12	410.74	461.02	511.40	561.82	12
14	412.41	462.70	513.08	563.50	14
16	414.08	464.37	514.76	565.18	16
18	415.75	466.05	516.44	566.86	18
20	417.43	467.73	518.12	568.54	20
22	419.10	469.41	519.80	570.22	22
24	420.77	471.08	521.48	571.90	24
26	422.45	472.76	523.16	573.58	26
28	424.12	474.43	524.85	575.27	28
30	425.79	476.10	526.53	576.95	30
32	427.47	477.78	528.21	578.63	32
34	429.15	479.46	529.89	580.32	34
36	430.82	481.14	531.57	582.00	36
38	432.50	482.83	533.25	583.69	38
40	434.18	484.51	534.93	585.37	40
42	435.86	486.19	536.61	587.05	42
44	437.54	487.87	538.29	588.74	44
46	439.21	489.56	539.97	590.42	46
48	440.89	491.24	541.65	592.11	48
50	442.57	492.92	543.33	593.79	50
52	444.25	494.60	545.01	595.47	52
54	445.93	496.28	546.69	597.16	54
56	447.60	497.96	548.37	598.84	56
58	449.28	499.65	550.06	600.53	58
60	450.95	501.32	551.74	602.22	60

′	12°	13°	14°	15°	′
0	602.22	652.87	703.53	754.35	0
2	603.91	654.56	705.23	756.05	2
4	605.60	656.25	706.92	757.74	4
6	607.28	657.93	708.62	759.44	6
8	608.97	659.62	710.31	761.13	8
10	610.66	661.31	712.01	762.83	10
12	612.35	663.00	713.71	764.53	12
14	614.04	664.69	715.40	766.22	14
16	615.72	666.37	717.10	767.92	16
18	617.41	668.06	718.79	769.61	18
20	619.10	669.75	720.49	771.31	20
22	620.79	671.44	722.20	773.01	22
24	622.48	673.13	723.89	774.70	24
26	624.16	674.81	725.59	776.40	26
28	625.85	676.51	727.28	778.09	28
30	627.55	678.20	728.97	779.79	30
32	629.24	679.89	730.66	781.49	32
34	630·93	681.58	732.35	783.19	34
36	632.61	683.26	734.05	784.89	36
38	634.30	684.95	735.74	786.59	38
40	635.99	686.64	737.43	788.29	40
42	637.68	688.33	739.12	789.99	42
44	639.37	690.02	740.81	791.69	44
46	641.05	691.70	742.51	793.39	46
48	642.74	693.39	744.20	795.09	48
50	644.43	695.08	745.89	796.79	50
52	646.12	696.77	747.58	798.49	52
54	647.81	698.46	749.27	800.19	54
56	649.49	700.14	750.97	801.89	56
58	651.18	701.83	752.66	803.59	58
60	652.87	703.53	754.35	805.29	60

′	16°	17°	18°	19°	′
0	805.29	856.35	907.52	958.86	0
2	806.99	858.05	909.23	960.57	2
4	808.64	859.76	910.94	962.30	4
6	810.39	861.46	912.65	964.00	6
8	812.09	863.16	914.36	965.72	8
10	813.79	864.87	916.07	967.43	10
12	815.49	866.57	917.78	969.15	12
14	817.19	868.27	919.49	970.86	14
16	818.89	869.98	921.20	972.58	16
18	820.59	871.68	922.91	974.29	18
20	822.29	873.38	924.63	976.01	20
22	823.99	875.09	926.34	977.72	22
24	825.69	876.79	928.05	979.44	24
26	827.39	878.49	929.76	981.15	26
28	829.09	880.20	931.47	982.86	28
30	830.79	881.90	933.18	984.58	30
32	832.49	883.61	934.89	986.30	32
34	834.20	885.32	936.60	988.02	34
36	835.90	887.02	938.32	989.74	36
38	837.61	888.73	940.03	991.46	38
40	839.31	890.44	941.74	993.18	40
42	841.01	892.15	943.45	994.90	42
44	842.72	893.86	945.16	996.62	44
46	844.42	895.56	946.88	998.34	46
48	846.13	897.27	948.59	1000.0	48
50	847.83	898.98	950.30	1001.8	50
52	849.53	900.69	952.01	1003.5	52
54	851.24	902.40	953.72	1005.2	54
56	852.94	904.10	955.44	1006.9	56
58	854.65	905.81	957.15	1008.6	58
60	856.35	907.52	958.86	1010.4	60

′	20°	21°	22°	23°	′
0	1010.4	1062.0	1113.8	1165.8	0
2	1012.1	1063.7	1115.5	1167.5	2
4	1013.8	1065.4	1117.3	1169.2	4
6	1015.5	1067.2	1119.0	1171.0	6
8	1017.2	1068.9	1120.7	1172.7	8
10	1019.0	1070.6	1122.4	1174.4	10
12	1020.7	1072.4	1124.2	1176.2	12
14	1022.4	1074.1	1125.9	1177.9	14
16	1024.1	1075.8	1127.6	1179.7	16
18	1025.8	1077.5	1129.4	1181.4	18
20	1027.6	1079.3	1131.1	1183.1	20
22	1029.3	1081.0	1132.8	1184.9	22
24	1031.0	1082.7	1134.6	1186.6	24
26	1032.7	1084.4	1136.3	1188.4	26
28	1034.4	1086.2	1138.0	1190.1	28
30	1036.1	1087.9	1139.7	1191.8	30
32	1037.9	1089.6	1141.5	1193.6	32
34	1039.6	1091.3	1143.2	1195.3	34
36	1041.3	1093.1	1144.9	1197.1	36
38	1043.0	1094.8	1146.7	1198.8	38
40	1044.8	1096.5	1148.4	1200.5	40
42	1046.5	1098.3	1150.1	1202.3	42
44	1048.2	1100.0	1151.9	1204.0	44
46	1049.9	1101.7	1153.6	1205.8	46
48	1051.7	1103.4	1155.4	1207.5	48
50	1053.4	1105.2	1157.1	1209.2	50
52	1055.1	1106.9	1158.8	1211.0	52
54	1056.8	1108.6	1160.6	1212.7	54
56	1058.6	1110.3	1162.3	1214.5	56
58	1060.3	1112.1	1164.0	1216.2	58
60	1062.0	1113.8	1165.8	1218.0	60

′	24°	25°	26°	27°	′
0	1218.0	1270.3	1322.9	1375.6	0
2	1219.7	1272.0	1324.6	1377.4	2
4	1221.4	1273.8	1326.4	1379.2	4
6	1223.2	1275.5	1328.1	1380.9	6
8	1224.9	1277.3	1329.9	1382.7	8
10	1226.7	1279.0	1331.6	1384.5	10
12	1228.4	1280.8	1333.4	1386.2	12
14	1230.2	1282.5	1335.2	1388.0	14
16	1231.9	1284.3	1336.9	1389.8	16
18	1233.6	1286.1	1338.7	1391.5	18
20	1235.4	1287.8	1340.4	1393.3	20
22	1237.1	1289.6	1342.2	1395.0	22
24	1238.9	1291.3	1343.9	1396.8	24
26	1240.6	1293.1	1345.7	1398.6	26
28	1242.4	1294.8	1347.4	1400.3	28
30	1244.1	1296.6	1349.2	1402.1	30
32	1245.8	1298.3	1351.0	1403.9	32
34	1247.6	1300.1	1352.7	1405.6	34
36	1249.3	1301.8	1354.5	1407.4	36
38	1251.1	1303.6	1356.2	1409.2	38
40	1252.8	1305.3	1358.0	1410.9	40
42	1254.6	1307.1	1359.8	1412.7	42
44	1256.3	1308.8	1361.5	1414.5	44
46	1258.1	1310.6	1363.3	1416.3	46
48	1259.8	1312.4	1365.1	1418.0	48
50	1261.5	1314.1	1366.8	1419.8	50
52	1263.3	1315.9	1368.6	1421.6	52
54	1265.0	1317.6	1370.4	1423.3	54
56	1266.8	1319.4	1372.1	1425.1	56
58	1268.5	1321.1	1373.9	1426.9	58
60	1270.3	1322.9	1375.6	1428.6	60

′	28°	29°	30°	31°	′
0	1428.6	1481.9	1535.3	1589.0	0
2	1430.4	1483.7	1537.1	1590.8	2
4	1432.2	1485.4	1538.9	1592.6	4
6	1434.0	1487.2	1540.7	1594.4	6
8	1435.7	1489.0	1542.5	1596.2	8
10	1437.5	1490.8	1544.3	1598.0	10
12	1439.3	1492.6	1546.0	1599.8	12
14	1441.1	1494.3	1547.8	1601.6	14
16	1442.8	1496.1	1549.6	1603.4	16
18	1444.6	1497.9	1551.4	1605.2	18
20	1446.4	1499.7	1553.2	1607.0	20
22	1448.2	1501.5	1555.0	1608.8	22
24	1449.9	1503.2	1556.8	1610.6	24
26	1451.7	1505.0	1558.6	1612.4	26
28	1453.5	1506.8	1560.4	1614.2	28
30	1455.2	1508.6	1562.2	1616.0	30
32	1457.0	1510.4	1564.0	1617.8	32
34	1458.8	1512.1	1565.7	1619.0	34
36	1460.6	1513.9	1567.5	1621.4	36
38	1462.3	1515.7	1569.3	1623.2	38
40	1464.1	1517.5	1571.1	1625.0	40
42	1465.9	1519.3	1572.9	1626.8	42
44	1467.7	1521.0	1574.7	1628.6	44
46	1469.5	1522.8	1576.5	1630.5	46
48	1471.2	1524.6	1578.3	1632.3	48
50	1473.0	1526.4	1580.1	1634.1	50
52	1474.8	1528.2	1581.9	1635.9	52
54	1476.6	1530.0	1583.7	1637.7	54
56	1478.3	1531.7	1585.5	1639.5	56
58	1480.1	1533.5	1587.2	1641.3	58
60	1481.9	1535.3	1589.0	1643.1	60

′	32°	33°	34°	35°	′
0	1643.1	1697.3	1751.8	1806.7	0
2	1644.9	1699.1	1753.7	1808.5	2
4	1646.7	1700.9	1755.5	1810.3	4
6	1648.5	1702.7	1757.3	1812.2	6
8	1650.3	1704.5	1759.1	1814.0	8
10	1652.1	1706.4	1761.0	1815.8	10
12	1653.9	1708.2	1762.8	1817.7	12
14	1655.7	1710.0	1764.6	1819.5	14
16	1657.5	1711.8	1766.4	1821.3	16
18	1659.3	1713.6	1768.3	1823.2	18
20	1661.1	1715.5	1770.1	1825.0	20
22	1662.9	1717.3	1771.9	1826.8	22
24	1664.7	1719.1	1773.7	1828.7	24
26	1666.5	1720.9	1775.6	1830.5	26
28	1668.3	1722.7	1777.4	1832.3	28
30	1670.1	1724.6	1779.2	1834.2	30
32	1671.9	1726.4	1781.0	1836.0	32
34	1673.7	1728.2	1782.9	1837.8	34
36	1675.5	1730.0	1784.7	1839.7	36
38	1677.4	1731.8	1786.5	1841.5	38
40	1679.2	1733.6	1788.4	1843.4	40
42	1681.0	1735.5	1790.2	1845.2	42
44	1682.8	1737.3	1792.0	1847.1	44
46	1684.6	1739.1	1793.9	1848.9	46
48	1686.4	1740.9	1795.7	1850.7	48
50	1688.2	1742.7	1797.5	1852.6	50
52	1690.0	1744.6	1799.3	1854.4	52
54	1691.8	1746.4	1801.2	1856.3	54
56	1693.7	1748.2	1803.0	1858.1	56
58	1695.5	1750.0	1804.8	1859.9	58
60	1697.3	1751.8	1806.7	1861.8	60

40

′	36°	37°	38°	39°	′
0	1861.8	1917.3	1973.0	2029.1	0
2	1863.6	1919.1	1974.9	2031.0	2
4	1865.5	1921.0	1976.7	2032.9	4
6	1867.3	1922.8	1978.6	2034.7	6
8	1869.2	1924.7	1980.5	2036.6	8
10	1871.0	1926.5	1982.3	2038.5	10
12	1872.9	1928.4	1984.2	2040.4	12
14	1874.7	1930.2	1986.1	2042.3	14
16	1876.5	1932.1	1987.9	2044.1	16
18	1878.4	1933.9	1989.8	2046.0	18
20	1880.2	1935.8	1991.7	2047.9	20
22	1882.1	1937.6	1993.6	2049.8	22
24	1883.9	1939.5	1995.4	2051.7	24
26	1885.8	1941.3	1997.3	2053.5	26
28	1887.6	1943.2	1999.2	2055.4	28
30	1889.5	1945.0	2001.0	2057.3	30
32	1891.3	1946.9	2002.9	2059.2	32
34	1893.2	1948.8	2004.8	2061.1	34
36	1895.0	1950.6	2006.6	2063.0	36
38	1896.9	1952.5	2008.5	2064.8	38
40	1898.7	1954.4	2010.4	2066.7	40
42	1900.6	1956.2	2012.3	2068.6	42
44	1902.4	1958.1	2014.1	2070.5	44
46	1904.3	1960.0	2016.0	2072.4	46
48	1906.1	1961.8	2017.9	2074.2	48
50	1908.0	1963.7	2019.7	2076.1	50
52	1909.8	1965.5	2021.6	2078.0	52
54	1911.7	1967.4	2023.5	2079.9	54
56	1913.5	1969.3	2025.4	2081.8	56
58	1915.4	1971.1	2027.2	2083.7	58
60	1917.3	1973.0	2029.1	2085.5	60

′	40°	41°	42°	43°	′
0	2085.5	2142.3	2199.5	2257.1	0
2	2087.4	2144.2	2201.4	2259.0	2
4	2089.3	2146.1	2203.3	2261.0	4
6	2091.2	2148.0	2205.3	2262.9	6
8	2093.1	2149.9	2207.2	2264.8	8
10	2095.0	2151.9	2209.1	2266.7	10
12	2096.9	2153.8	2211.0	2268.7	12
14	2098.8	2155.7	2212.9	2270.6	14
16	2100.7	2157.6	2214.9	2272.5	16
18	2102.6	2159.5	2216.8	2274.5	18
20	2104.5	2161.4	2218.7	2276.4	20
22	2106.3	2163.3	2220.6	2278.3	22
24	2108.2	2165.2	2222.5	2280.2	24
26	2110.1	2167.1	2224.4	2282.2	26
28	2112.0	2169.0	2226.4	2284.1	28
30	2113.9	2170.9	2228.3	2286.0	30
32	2115.8	2172.8	2230.2	2288.0	32
34	2117.7	2174.7	2232.1	2289.9	34
36	2119.6	2176.6	2234.0	2291.8	36
38	2121.5	2178.5	2236.0	2293.8	38
40	2123.4	2180.4	2237.9	2295.7	40
42	2125.3	2182.4	2239.8	2297.7	42
44	2127.2	2184.3	2241.7	2299.6	44
46	2129.1	2186.2	2243.6	2301.5	46
48	2131.0	2188.1	2245.6	2303.5	48
50	2132.9	2190.0	2247.5	2305.4	50
52	2134.7	2191.9	2249.4	2307.3	52
54	2136.6	2193.8	2251.3	2309.3	54
56	2138.5	2195.7	2253.3	2311.2	56
58	2140.4	2197.6	2255.2	2313.1	58
60	2142.3	2199.5	2257.1	2315.1	60

′	44°	45°	46°	47°	′
0	2315.1	2373.4	2432.2	2491.5	0
2	2317.0	2375.4	2434.2	2493.4	2
4	2319.0	2377.3	2436.1	2495.4	4
6	2320.9	2379.3	2438.1	2497.4	6
8	2322.8	2381.2	2440.1	2499.4	8
10	2324.8	2383.2	2442.1	2501.4	10
12	2326.7	2385.2	2444.0	2503.4	12
14	2328.7	2387.1	2446.0	2505.4	14
16	2330.6	2389.1	2448.0	2507.3	16
18	2332.6	2391.0	2449.9	2509.3	18
20	2334.5	2393.0	2451.9	2511.3	20
22	2336.4	2394.9	2453.9	2513.3	22
24	2338.4	2396.9	2455.9	2515.3	24
26	2340.3	2398.8	2457.8	2517.3	26
28	2342.3	2400.8	2459.8	2519.3	28
30	2344.2	2402.8	2461.8	2521.2	30
32	2346.1	2404.7	2463.8	2523.2	32
34	2348.1	2406.7	2465.7	2525.2	34
36	2350.0	2408.6	2467.7	2527.2	36
38	2352.0	2410.6	2469.7	2529.2	38
40	2353.9	2412.6	2471.7	2531.2	40
42	2355.9	2414.5	2473.6	2533.2	42
44	2357.8	2416.5	2475.6	2535.2	44
46	2359.8	2418.5	2477.6	2537.2	46
48	2361.7	2420.4	2479.6	2539.2	48
50	2363.7	2422.4	2481.6	2541.2	50
52	2365.6	2424.4	2483.5	2543.1	52
54	2367.6	2426.3	2485.5	2545.1	54
56	2369.5	2428.3	2487.5	2547.1	56
58	2371.5	2430.2	2489.5	2549.1	58
60	2373.4	2432.2	2491.5	2551.1	60

′	48°	49°	50°	51°	′
0	2551.1	2611.3	2671.9	2733.0	0
2	2553.1	2613.3	2673.9	2735.1	2
4	2555.1	2615.3	2676.0	2737.1	4
6	2557.1	2617.3	2678.0	2739.2	6
8	2559.1	2619.3	2680.0	2741.2	8
10	2561.1	2621.4	2682.1	2743.3	10
12	2563.1	2623.4	2684.1	2745.3	12
14	2565.1	2625.4	2686.1	2747.4	14
16	2567.1	2627.4	2688.2	2749.4	16
18	2569.1	2629.4	2690.2	2751.5	18
20	2571.1	2631.4	2692.3	2753.5	20
22	2573.1	2633.5	2694.3	2755.6	22
24	2575.1	2635.5	2696.3	2757.7	24
26	2577.1	2637.5	2698.4	2759.7	26
28	2579.1	2639.5	2700.4	2761.8	28
30	2581.1	2641.5	2702.4	2763.8	30
32	2583.1	2643.5	2704.5	2765.9	32
34	2585.1	2645.6	2706.5	2767.9	34
36	2587.2	2647.6	2708.6	2770.0	36
38	2589.2	2649.6	2710.6	2772.0	38
40	2591.2	2651.6	2712.6	2774.1	40
42	2593.2	2653.7	2714.7	2776.2	42
44	2595.2	2655.7	2716.7	2778.2	44
46	2597.2	2657.7	2718.8	2780.3	46
48	2599.2	2659.7	2720.8	2782.3	48
50	2601.2	2661.8	2722.8	2784.4	50
52	2603.2	2663.8	2724.9	2786.4	52
54	2605.2	2665.8	2726.9	2788.5	54
56	2607.2	2667.8	2729.0	2790.6	56
58	2609.3	2669.9	2731.0	2792.6	58
60	2611.3	2671.9	2733.0	2794.7	60

′	52°	53°	54°	55°	′
0	2794.7	2856.9	2919.5	2982.8	0
2	2796.8	2858.9	2921.6	2984.9	2
4	2798.8	2861.0	2923.8	2987.1	4
6	2800.9	2863.1	2925.9	2989.2	6
8	2803.0	2865.2	2928.0	2991.3	8
10	2805.0	2867.3	2930.1	2993.4	10
12	2807.1	2869.4	2932.2	2995.5	12
14	2809.2	2871.5	2934.3	2997.7	14
16	2811.2	2873.5	2936.4	2999.8	16
18	2813.3	2875.6	2938.5	3001.9	18
20	2815.4	2877.7	2940.6	3004.0	20
22	2817.4	2879.8	2942.7	3006.2	22
24	2819.5	2881.9	2944.8	3008.3	24
26	2821.6	2884.0	2946.9	3010.4	26
28	2823.6	2886.1	2949.0	3012.5	28
30	2825.7	2888.1	2951.1	3014.7	30
32	2827.8	2890.2	2953.2	3016.8	32
34	2829.8	2892.3	2955.3	3018.9	34
36	2831.9	2894.4	2957.5	3021.1	36
38	2834.0	2896.5	2959.6	3023.2	38
40	2836.1	2898.6	2961.7	3025.3	40
42	2838.2	2900.7	2963.8	3027.5	42
44	2840.2	2902.8	2965.9	3029.6	44
46	2842.3	2904.9	2968.0	3031.7	46
48	2844.4	2907.0	2970.1	3033.8	48
50	2846.5	2909.1	2972.2	3036.0	50
52	2848.5	2911.2	2974.4	3038.1	52
54	2850.6	2913.3	2976.5	3040.2	54
56	2852.7	2915.4	2978.6	3042.4	56
58	2854.8	2917.5	2980.7	3044.5	58
60	2856.9	2919.5	2982.8	3046.6	60

′	56°	57°	58°	59°	′
0	3046.6	3111.1	3176.1	3241.9	0
2	3048.8	3113.3	3178.3	3244.1	2
4	3050.9	3115.4	3180.5	3246.3	4
6	3053.1	3117.6	3182.7	3248.5	6
8	3055.2	3119.7	3184.9	3250.7	8
10	3057.4	3121.9	3187.1	3252.9	10
12	3059.5	3124.1	3189.2	3255.1	12
14	3061.6	3126.2	3191.4	3257.3	14
16	3063.8	3128.4	3193.6	3259.5	16
18	3065.9	3130.6	3195.8	3261.7	18
20	3068.1	3132.7	3198.0	3263.9	20
22	3070.2	3134.9	3200.2	3266.1	22
24	3072.4	3137.0	3202.4	3268.3	24
26	3074.5	3139.2	3204.5	3270.5	26
28	3076.6	3141.4	3206.7	3272.7	28
30	3078.8	3143.5	3208.9	3274.9	30
32	3080.9	3145.7	3211.1	3277.1	32
34	3083.1	3147·9	3213.3	3279.4	34
36	3085.2	3150.0	3215.5	3281.6	36
38	3087.4	3152.2	3217.7	3283.8	38
40	3089.6	3154.4	3219.9	3286.0	40
42	3091.7	3156.6	3222.1	3288.2	42
44	3093.9	3158.7	3224.3	3290.5	44
46	3096.0	3160.9	3226.5	3292.7	46
48	3098.2	3163.1	3228.7	3294.9	48
50	3100.3	3165.3	3230.9	3297.1	50
52	3102.5	3167.4	3233.1	3299.3	52
54	3104.6	3169.6	3235.3	3301.5	54
56	3106.8	3171.8	3237.5	3303.8	56
58	3108.9	3174.0	3239.7	3306.0	58
60	3111.1	3176.1	3241.9	3308.2	60

′	60°	61°	62°	63°	′
0	3308.2	3375.2	3442.9	3511.3	0
2	3310.4	3377.4	3445.2	3513.6	2
4	3312.7	3379.7	3447.5	3515.9	4
6	3314.9	3381.9	3449.7	3518.2	6
8	3317.1	3384.2	3452.0	3520.5	8
10	3319.3	3386.4	3454.3	3522.8	10
12	3321.6	3388.7	3456.6	3525.1	12
14	3323.8	3390.9	3458.8	3527.4	14
16	3326.0	3393.2	3461.1	3529.7	16
18	3328.3	3395.4	3463.4	3532.0	18
20	3330.5	3397.7	3465.7	3534.3	20
22	3332.7	3399.9	3467.9	3536.6	22
24	3334.9	3402.2	3470.2	3538.9	24
26	3337.2	3404.4	3472.5	3541.2	26
28	3339.4	3406.7	3474.7	3543.5	28
30	3341.6	3408.9	3477.0	3545.8	30
32	3343.9	3411.2	3479.3	3548.1	32
34	3346.1	3413.5	3481.6	3550.4	34
36	3348.3	3415.7	3483.9	3552.7	36
38	3350.6	3418.0	3486.2	3555.0	38
40	3352.8	3420.3	3488.5	3557.3	40
42	3355.0	3422.5	3490.7	3559.6	42
44	3357.3	3424.8	3493.0	3562.0	44
46	3359.5	3427.1	3495.3	3564.3	46
48	3361.8	3429.3	3497.6	3566.6	48
50	3364.0	3431.6	3499.9	3568.9	50
52	3366.2	3433.9	3502.2	3571.2	52
54	3368.5	3436.1	3504.5	3573.5	54
56	3370.7	3438.4	3506.8	3575.8	56
58	3373.0	3440.7	3509.0	3578.1	58
60	3375.2	3442.9	3511.3	3580.4	60

′	64°	65°	66°	67°	′
0	3580.4	3650.4	3721.1	3792.6	0
2	3582.8	3652.8	3723.4	3795.0	2
4	3585.1	3655.1	3725.8	3797.4	4
6	3587.4	3657.5	3728.2	3799.8	6
8	3589.7	3659.8	3730.6	3802.2	8
10	3592.1	3662.2	3732.9	3804.6	10
12	3594.4	3664.5	3735.3	3807.0	12
14	3596.7	3666.9	3737.7	3809.4	14
16	3599.1	3669.2	3740.1	3811.8	16
18	3601.4	3671.6	3742.4	3814.2	18
20	3603.7	3673.9	3744.8	3816.6	20
22	3606.0	3676.2	3747.2	3819.0	22
24	3608.4	3678.6	3749.6	3821.4	24
26	3610.7	3680.9	3751.9	3823.8	26
28	3613.0	3683.3	3754.3	3826.2	28
30	3615.3	3685.6	3756.7	3828.6	30
32	3617.7	3688.0	3759.1	3831.0	32
34	3620.0	3690.4	3761.5	3833.4	34
36	3622.3	3692.7	3763.9	3835.9	36
38	3624.7	3695.1	3766.3	3838.3	38
40	3627.0	3697.4	3768.7	3840.7	40
42	3629.4	3699.8	3771.0	3843.1	42
44	3631.7	3702.2	3773.4	3845.5	44
46	3634.0	3704.5	3775.8	3847.9	46
48	3636.4	3706.9	3778.2	3850.4	48
50	3638.7	3709.3	3780.6	3852.8	50
52	3641.1	3711.6	3783.0	3855.2	52
54	3643.4	3714.0	3785.4	3857.6	54
56	3645.7	3716.3	3787.8	3860.0	56
58	3648.1	3718.7	3790.2	3862.5	58
60	3650.4	3721.1	3792.6	3864.9	60

′	68°	69°	70°	71°	′
0	3864.9	3938.1	4012.1	4087.1	0
2	3867.3	3940.6	4014.6	4089.7	2
4	3869.7	3943.0	4017.1	4092.2	4
6	3872.2	3945.5	4019.6	4094.7	6
8	3874.6	3947.9	4022.1	4097.2	8
10	3877.0	3950.4	4024.6	4099.8	10
12	3879.5	3952.9	4027.1	4102.3	12
14	3881.9	3955.3	4029.6	4104.8	14
16	3884.3	3957.8	4032.1	4107.3	16
18	3886.8	3960.2	4034.6	4109.8	18
20	3889.2	3962.7	4037.1	4112.4	20
22	3891.6	3965.2	4039.6	4114.9	22
24	3894.1	3967.6	4042.1	4117.4	24
26	3896.5	3970.1	4044.6	4119.9	26
28	3898.9	3972.5	4047.1	4122.4	28
30	3901.4	3975.0	4049.6	4125.0	30
32	3903.8	3977.5	4052.1	4127.5	32
34	3906.3	3980.0	4054.6	4130.0	34
36	3908.7	3982.4	4057.1	4132.6	36
38	3911.2	3984.9	4059.6	4135.1	38
40	3913.6	3987.4	4062.1	4137.7	40
42	3916.1	3989.9	4064.6	4140.2	42
44	3918.5	3992.3	4067.1	4142.7	44
46	3921.0	3994.8	4069.6	4145.3	46
48	3923.4	3997.3	4072.1	4147.8	48
50	3925.9	3999.8	4074.6	4150.4	50
52	3928.3	4002.2	4077.1	4152.9	52
54	3930.8	4004.7	4079.6	4155.4	54
56	3933.2	4007.2	4082.1	4158.0	56
58	3935.7	4009.7	4084.6	4160.5	58
60	3938.1	4012.1	4087.1	4163.1	60

′	72°	73°	74°	75°	′
0	4163.1	4240.0	4317.8	4396.7	0
2	4165.6	4242.6	4320.5	4399.4	2
4	4168.2	4245.1	4323.1	4402.1	4
6	4170.7	4247.7	4325.7	4404.7	6
8	4173.3	4250.3	4328.3	4407.4	8
10	4175.8	4252.9	4330.9	4410.0	10
12	4178.4	4255.5	4333.6	4412.7	12
14	4181.0	4258.1	4336.2	4415.3	14
16	4183.5	4260.7	4338.8	4418.0	16
18	4186.1	4263.2	4341.4	4420.7	18
20	4188.6	4265.8	4344.0	4423.3	20
22	4191.2	4268.4	4346.7	4426.0	22
24	4193.7	4271.0	4349.3	4428.6	24
26	4196.3	4273.6	4351.9	4431.3	26
28	4198.8	4276.2	4354.5	4434.0	28
30	4201.4	4278.8	4357.1	4436.6	30
32	4204.0	4281.4	4359.8	4439.3	32
34	4206.5	4284.0	4362.4	4442.0	34
36	4209.1	4286.6	4365.1	4444.6	36
38	4211.7	4289.2	4367.7	4447.3	38
40	4214.3	4291.8	4370.3	4450.0	40
42	4216.8	4294.4	4373.0	4452.7	42
44	4219.4	4297.0	4375.6	4455.3	44
46	4222.0	4299.6	4378.3	4458.0	46
48	4224.5	4302.2	4380.9	4460.7	48
50	4227.1	4304.8	4383.5	4463.4	50
52	4229.7	4307.4	4386.2	4466.0	52
54	4232.3	4310.0	4388.8	4468.7	54
56	4234.8	4312.6	4391.5	4471.4	56
58	4237.4	4315.2	4394.1	4474.1	58
60	4240.0	4317.8	4396.7	4476.7	60

′	76°	77°	78°	79°	′
0	4476.7	4557.8	4640.0	4723.4	0
2	4479.4	4560.5	4642.8	4726.2	2
4	4482.1	4563.3	4645.6	4729.0	4
6	4484.8	4566.0	4648.3	4731.8	6
8	4487.5	4568.7	4651.1	4734.7	8
10	4490.2	4571.5	4653.9	4737.5	10
12	4492.9	4574.2	4656.7	4740.3	12
14	4495.6	4576.9	4659.4	4743.1	14
16	4498.3	4579.7	4662.2	4745.9	16
18	4501.0	4582.4	4665.0	4748.7	18
20	4503.7	4585.1	4667.7	4751.5	20
22	4506.3	4587.9	4670.5	4754.3	22
24	4509.0	4590.6	4673.3	4757.1	24
26	4511.7	4593.3	4676.0	4760.0	26
28	4514.4	4596.0	4678.8	4762.8	28
30	4517.1	4598.8	4681.6	4765.6	30
32	4519.8	4601.5	4684.4	4768.4	32
34	4522.5	4604.3	4687.2	4771.2	34
36	4525.3	4607.0	4689.9	4774.1	36
38	4528.0	4609.8	4692.7	4776.9	38
40	4530.7	4612.5	4695.5	4779.7	40
42	4533.4	4615.3	4698.3	4782.6	42
44	4536.1	4618.0	4701.1	4785.4	44
46	4538.8	4620.8	4703.9	4788.2	46
48	4541.5	4623.5	4706.7	4791.0	48
50	4544.2	4626.3	4709.5	4793.9	50
52	4547.0	4629.0	4712.2	4796.7	52
54	4549.7	4631.8	4715.0	4799.5	54
56	4552.4	4634.5	4717.8	4802.4	56
58	4555.1	4637.3	4720.6	4805.2	58
60	4557.8	4640.0	4723.4	4808.0	60

′	80°	81°	82°	83°	′
0	4808.0	4893.9	4981.0	5069.4	0
2	4810.9	4896.8	4983.9	5072.4	2
4	4813.7	4899.7	4986.8	5075.4	4
6	4816.6	4902.6	4989.8	5078.4	6
8	4819.4	4905.4	4992.7	5081.4	8
10	4822.3	4908.3	4995.7	5084.4	10
12	4825.1	4911.2	4998.6	5087.3	12
14	4828.0	4914.1	5001.5	5090.3	14
16	4830.8	4917.0	5004.5	5093.3	16
18	4833.7	4919.9	5007.4	5096.3	18
20	4836.5	4922.8	5010.3	5099.3	20
22	4839.4	4925.7	5013.3	5102.3	22
24	4842.2	4928.6	5016.2	5105.2	24
26	4845.1	4931.5	5019.2	5108.2	26
28	4847.9	4934.4	5022.1	5111.2	28
30	4850.8	4937.2	5025.0	5114.2	30
32	4853.7	4940.2	5028.0	5117.2	32
34	4856.5	4943.1	5031.0	5120.2	34
36	4859.4	4946.0	5033.9	5123.2	36
38	4862.3	4948.9	5036.9	5126.2	38
40	4865.1	4951.8	5039.8	5129.2	40
42	4868.0	4954.7	5042.8	5132.2	42
44	4870.9	4957.6	5045.8	5135.2	44
46	4873.8	4960.6	5048.7	5138.2	46
48	4876.6	4963.5	5051.7	5141.2	48
50	4879.5	4966.4	5054.6	5144.3	50
52	4882.4	4969.3	5057.6	5147.3	52
54	4885.3	4972.2	5060.6	5150.3	54
56	4888.1	4975.1	5063.5	5153.3	56
58	4891.0	4978.0	5066.5	5156.3	58
60	4893.9	4981.0	5069.4	5159.3	60

′	84°	85°	86°	87°	′
0	5159.3	5250.6	5343.3	5437.5	0
2	5162.3	5253.6	5346.4	5440.7	2
4	5165.3	5256.7	5349.5	5443.9	4
6	5168.4	5259.8	5352.7	5447.1	6
8	5171.4	5262.9	5355.8	5450.3	8
10	5174.4	5266.0	5358.9	5453.4	10
12	5177.5	5269.0	5362.0	5456.6	12
14	5180.5	5272.1	5365.2	5459.8	14
16	5183.5	5275.2	5368.3	5463.0	16
18	5186.6	5278.3	5371.4	5466.2	18
20	5189.6	5281.4	5374.6	5469.4	20
22	5192.6	5284.4	5377.7	5472.5	22
24	5195.6	5287.5	5380.8	5475.7	24
26	5198.7	5290.6	5383.9	5478.9	26
28	5201.7	5293.7	5387.1	5482.1	28
30	5204.7	5296.7	5390.2	5485.3	30
32	5207.8	5299.8	5393.4	5488.5	32
34	5210.8	5302.9	5396.5	5491.7	34
36	5213.9	5306.1	5399.7	5494.9	36
38	5216.9	5309.2	5402.8	5498.1	38
40	5220.0	5312.3	5406.0	5501.3	40
42	5223.1	5315.4	5409.1	5504.5	42
44	5226.1	5318.5	5412.3	5507.7	44
46	5229.2	5321.6	5415.4	5510.9	46
48	5232.2	5324.7	5418.6	5514.1	48
50	5235.3	5327.8	5421.8	5517.3	50
52	5238.3	5330.9	5424.9	5520.5	52
54	5241.4	5334.0	5428.1	5523.7	54
56	5244.5	5337.1	5431.2	5526.9	56
58	5247.5	5340.2	5434.4	5530.1	58
60	5250.6	5343.3	5437.5	5533.3	60

53

′	88°	′	88°	′	89°	′	89°
0	5533.3	30	5581.9	0	5630.8	30	5680.2
2	5536.6	32	5585.1	2	5634.1	32	5683.5
4	5539.8	34	5588.4	4	5637.4	34	5686.8
6	5543.1	36	5591.7	6	5640.7	36	5690.2
8	5546.3	38	5594.9	8	5644.0	38	5693.5
10	5549.5	40	5598.2	10	5647.3	40	5696.8
12	5552.8	42	5601.4	12	5650.6	42	5700.1
14	5556.0	44	5604.7	14	5653.9	44	5703.4
16	5559.2	46	5608.0	16	5657.1	46	5706.8
18	5562.5	48	5611.2	18	5660.4	48	5710.1
20	5565.7	50	5614.5	20	5663.7	50	5713.4
22	5568.9	52	5617.8	22	5667.0	52	5716.7
24	5572.2	54	5621.0	24	5670.3	54	5720.0
26	5575.4	56	5624.3	26	5673.6	56	5723.4
28	5578.6	58	5627.5	28	5676.9	58	5726.7
30	5581.9	60	5630.8	30	5680.2	60	5730.0

TABLE

GIVING RADII OF DEGREES OF CURVE.

RADII OF CURVES.

0°

'	RADIUS.	'	RADIUS.	'	RADIUS.	'	RADIUS.
0	INFINITE.	15	22918.3	30	11459.2	45	7639.49
1	343775.	16	21485.9	31	11089.6	46	7473.42
2	171887.	17	20222.1	32	10743.0	47	7314.41
3	114592.	18	19098.6	33	10417.5	48	7162.03
4	85943.7	19	18093.4	34	10111.1	49	7015.87
5	68754.9	20	17188.8	35	9822.18	50	6875.55
6	57295.8	21	16370.2	36	9549.34	51	6740.74
7	49110.7	22	15626.1	37	9291.29	52	6611.12
8	42971.8	23	14946.7	38	9046.75	53	6486.38
9	38197.2	24	14323.6	39	8814.78	54	6366.26
10	34377.5	25	13751.0	40	8594.41	55	6250.51
11	31252.3	26	13222.1	41	8384.80	56	6138.90
12	28647.8	27	12732.4	42	8185.16	57	6031.20
13	26444.2	28	12277.7	43	7994.81	58	5927.22
14	24555.4	29	11854.3	44	7813.11	59	5826.76
15	22918.3	30	11459.2	45	7639.49	60	5729.65

RADII OF CURVES.

1°

'	RADIUS.	'	RADIUS.	'	RADIUS.	'	RADIUS.
0	5729.65	15	4583.75	30	3819.83	45	3274.17
1	5635.72	16	4523.44	31	3777.85	46	3243.29
2	5544.83	17	4464.70	32	3736.79	47	3212.98
3	5456.82	18	4407.46	33	3696.61	48	3183.23
4	5371.56	19	4351.67	34	3657.29	49	3154.03
5	5288.92	20	4297.28	35	3618.80	50	3125.36
6	5208.79	21	4244.23	36	3581.10	51	3097.20
7	5131.05	22	4192.47	37	3544.19	52	3069.55
8	5055.59	23	4141.96	38	3508.02	53	3042.39
9	4982.33	24	4092.66	39	3472.59	54	3015.71
10	4911.15	25	4044.51	40	3437.87	55	2989.48
11	4841.98	26	3997.49	41	3403.83	56	2963.71
12	4774.74	27	3951.54	42	3370.46	57	2938.39
13	4709.33	28	3906.64	43	3337.74	58	2913.49
14	4645.69	29	3862.74	44	3305.65	59	2889.01
15	4583.75	30	3819.83	45	3274.17	60	2864.93

RADII OF CURVES.

2°

'	RADIUS.	'	RADIUS.	'	RADIUS.	'	RADIUS.
0	2864.93	15	2546.64	30	2292.01	45	2083.68
1	2841.26	16	2527.92	31	2276.84	46	2071.13
2	2817.97	17	2509.47	32	2261.86	47	2058.73
3	2795.06	18	2491.29	33	2247.08	48	2046.48
4	2772.53	19	2473.37	34	2232.49	49	2034.37
5	2750.35	20	2455.70	35	2218.09	50	2022.41
6	2728.52	21	2438.29	36	2203.87	51	2010.59
7	2707.04	22	2421.12	37	2189.84	52	1998.90
8	2685.89	23	2404.19	38	2175.98	53	1987.35
9	2665.08	24	2387.50	39	2162.30	54	1975.93
10	2644.58	25	2371.04	40	2148.79	55	1964.64
11	2624.39	26	2354.80	41	2135.44	56	1953.48
12	2604.51	27	2338.78	42	2122.26	57	1942.44
13	2584.93	28	2322.98	43	2109.24	58	1931.53
14	2565.65	29	2307.39	44	2096.39	59	1920.75
15	2546.64	30	2292.01	45	2083.68	60	1910.08

RADII OF CURVES.

3°

'	RADIUS.	'	RADIUS.	'	RADIUS.	'	RADIUS.
0	1910.08	15	1763.18	30	1637.28	45	1528.16
1	1899.53	16	1754.19	31	1629.52	46	1521.40
2	1889.09	17	1745.29	32	1621.84	47	1514.70
3	1878.77	18	1736.48	33	1614.22	48	1508.06
4	1868.56	19	1727.75	34	1606.68	49	1501.48
5	1858.47	20	1719.12	35	1599.21	50	1494.95
6	1848.48	21	1710.56	36	1591.81	51	1488.48
7	1838.59	22	1702.10	37	1584.48	52	1482.07
8	1828.82	23	1693.72	38	1577.21	53	1475.71
9	1819.14	24	1685.42	39	1570.01	54	1469.41
10	1809.57	25	1677.20	40	1562.88	55	1463.16
11	1800.10	26	1669.06	41	1555.81	56	1456.96
12	1790.73	27	1661.00	42	1548.80	57	1450.81
13	1781.45	28	1653.01	43	1541.86	58	1444.72
14	1772.27	29	1645.11	44	1534.98	59	1438.68
15	1763.18	30	1637.28	45	1528.16	60	1432.69

RADII OF CURVES.

4°

′	RADIUS.	′	RADIUS.	′	RADIUS.	′	RADIUS.
0	1432.69	15	1348.45	30	1273.57	45	1206.57
1	1426.74	16	1343.18	31	1268.87	46	1202.36
2	1420.85	17	1337.96	32	1264.21	47	1198.17
3	1415.01	18	1332.77	33	1259.58	48	1194.01
4	1409.21	19	13_7.63	34	1254.98	49	1189.88
5	1403.46	20	1322.53	35	1250.42	50	1185.78
6	1397.76	21	1317.46	36	1245.89	51	1181.71
7	1392.10	22	1312.43	37	1241.40	52	1177.66
8	1386.49	23	1307.45	38	1236.94	53	1173.65
9	1380.92	24	1302.50	39	1232.51	54	1169.66
10	1375.40	25	1297.58	40	1228.11	55	1165.70
11	1369.92	26	1292.71	41	1223.74	56	1161.76
12	1364.49	27	1287.87	42	1219.40	57	1157.85
13	1359.10	28	1283.07	43	1215.09	58	1153.97
14	1353.75	29	1278.30	44	1210.82	59	1150.11
15	1348.45	30	1273.57	45	1206.57	60	1146.28

RADII OF CURVES.

5°

'	RADIUS.	'	RADIUS.	'	RADIUS.	'	RADIUS.
0	1146.28	15	1091.73	30	1042.14	45	996.87
1	1142.47	16	1088.28	31	1039.00	46	993.99
2	1138.69	17	1084.85	32	1035.87	47	991.13
3	1134.94	18	1081.44	33	1032.76	48	988.28
4	1131.21	19	1078.05	34	1029.67	49	985.45
5	1127.50	20	1074.68	35	1026.60	50	982.64
6	1123.82	21	1071.34	36	1023.55	51	979.84
7	1120.16	22	1068.01	37	1020.51	52	977.06
8	1116.52	23	1064.71	38	1017.49	53	974.29
9	1112.91	24	1061.43	39	1014.50	54	971.54
10	1109.33	25	1058.16	40	1011.51	55	968.81
11	1105.76	26	1054.92	41	1008.55	56	966.09
12	1102.22	27	1051.70	42	1005.60	57	963.39
13	1098.70	28	1048.48	43	1002.67	58	960.70
14	1095.20	29	1045.31	44	999.76	59	958.02
15	1091.73	30	1042.14	45	996.87	60	955.37

RADII OF CURVES.

6°

′	RADIUS.	′	RADIUS.	′	RADIUS.	′	RADIUS.
0	955.37	15	917.19	30	881.95	45	849.32
1	952.72	16	914.75	31	879.69	46	847.23
2	950.09	17	912.33	32	877.45	47	845.15
3	947.48	18	909.92	33	875.22	48	843.08
4	944.88	19	907.52	34	873.00	49	841.02
5	942.29	20	905.13	35	870.79	50	838.97
6	939.72	21	902.76	36	868.60	51	836.93
7	937.16	22	900.40	37	866.41	52	834.90
8	934.62	23	898.05	38	864.24	53	832.89
9	932.09	24	895.71	39	862.07	54	830.88
10	929.57	25	893.39	40	859.92	55	828.88
11	927.07	26	891.08	41	857.78	56	826.89
12	924.58	27	888.78	42	855.65	57	824.91
13	922.10	28	886.49	43	853.53	58	822.93
14	919.64	29	884.21	44	851.42	59	820.97
15	917.19	30	881.95	45	849.32	60	819.02

RADII OF CURVES.

7°

'	RADIUS.	'	RADIUS.	'	RADIUS.	'	RADIUS.
0	819.02	15	790.81	30	764.49	45	739.86
1	817.08	16	789.00	31	762.80	46	738.28
2	815.14	17	787.96	32	761.11	47	736.70
3	813.22	18	785.40	33	759.43	48	735.13
4	811.30	19	783.62	34	757.76	49	733.56
5	809.40	20	781.84	35	756.10	50	732.01
6	807.50	21	780.07	36	754.45	51	730.45
7	805.61	22	778.31	37	752.80	52	728.91
8	803.73	23	776.55	38	751.16	53	727.37
9	801.86	24	774.81	39	749.52	54	725.84
10	800.00	25	773.07	40	747.89	55	724.31
11	798.14	26	771.34	41	746.27	56	722.79
12	796.30	27	769.61	42	744.66	57	721.28
13	794.46	28	767.90	43	743.06	58	719.77
14	792.63	29	766.19	44	741.46	59	718.27
15	790.81	30	764.49	45	739.86	60	716.78

RADII OF CURVES.

8°

'	RADIUS.	'	RADIUS.	'	RADIUS.	'	RADIUS.
0	716.78	15	695.09	30	674.69	45	655.45
1	715.29	16	693.70	31	673.37	46	654.20
2	713.81	17	692.30	32	672.06	47	652.96
3	712.33	18	690.91	33	670.75	48	651.73
4	710.87	19	689.53	34	669.45	49	650.50
5	709.40	20	688.16	35	668.15	50	649.27
6	707.95	21	686.78	36	666.86	51	648.05
7	706.49	22	685.42	37	665.57	52	616.84
8	705.05	23	684.06	38	664.29	53	645.63
9	703.61	24	682.70	39	663.01	54	644.42
10	702.18	25	681.35	40	661.74	55	643.22
11	700.75	26	680.01	41	660.47	56	642.02
12	699.33	27	678.67	42	659.21	57	640.83
13	697.91	28	677.34	43	657.95	58	639.64
14	696.50	29	676.01	44	556.69	59	638.46
15	695.09	30	674.69	45	655.45	60	637.28

RADII OF CURVES.

9°

'	RADIUS.	'	RADIUS.	'	RADIUS.	'	RADIUS.
0	637.28	15	620.09	30	603.81	45	588.36
1	636.10	16	618.97	31	602.75	46	587.36
2	634.93	17	617.87	32	601.70	47	586.36
3	633.76	18	616.76	33	600.65	48	585.36
4	632.60	19	615.66	34	599.61	49	584.37
5	631.44	20	614.56	35	598.57	50	583.39
6	630.29	21	613.47	36	597.53	51	582.40
7	629.14	22	612.38	37	596.50	52	581.42
8	627.99	23	611.30	38	595.47	53	580.44
9	626.85	24	610.21	39	594.44	54	579.47
10	625.71	25	609.14	40	593.42	55	578.49
11	624.58	26	608.06	41	592.40	56	577.53
12	623.45	27	606.99	42	591.38	57	576.56
13	622.33	28	605.93	43	590.37	58	575.60
14	621.20	29	604.86	44	589.36	59	574.64
15	620.09	30	603.81	45	588.36	60	573.69

RADII OF CURVES.

10°

'	RADIUS.	'	RADIUS.	'	RADIUS.	'	RADIUS.
0	573.69	15	559.73	30	546.41	45	533.77
1	572.73	16	558.82	31	545.58	46	532.94
2	571.78	17	557.92	32	544.71	47	532.12
3	570.84	18	557.02	33	543.86	48	531.30
4	569.90	19	556.12	34	543.00	49	530.49
5	568.96	20	555.23	35	542.15	50	529.67
6	568.02	21	554.24	36	541.30	51	528.86
7	567.09	22	553.45	37	540.45	52	528.05
8	566.16	23	552.56	38	539.61	53	527.25
9	565.23	24	551.68	39	538.76	54	526.44
10	564.30	25	550.80	40	537.92	55	525.64
11	563.38	26	549.92	41	537.09	56	524.84
12	562.47	27	549.05	42	536.25	57	524.05
13	561.55	28	548.17	43	535.42	58	523.25
14	560.64	29	547.31	44	534.59	59	522.46
15	559.73	30	546.44	45	533.77	60	521.67

RADII OF CURVES.

11°

'	RADIUS.	'	RADIUS.	'	RADIUS.	'	RADIUS.
0	521.67	15	510.12	30	499.06	45	488.48
1	520.89	16	509.36	31	498.34	46	487.79
2	520.10	17	508.61	32	497.62	47	487.10
3	519.32	18	507.87	33	496.91	48	486.42
4	518.54	19	507.12	34	496.19	49	485.73
5	517.76	20	506.38	35	495.48	50	485.05
6	516.99	21	505.63	36	494.77	51	484.37
7	516.21	22	504.90	37	494.07	52	483.69
8	515.44	23	504.16	38	493.36	53	483.02
9	514.68	24	503.42	39	492.66	54	482.34
10	513.91	25	502.69	40	491.96	55	481.67
11	513.15	26	501.96	41	491.26	56	481.00
12	512.38	27	501.23	42	490.56	57	480.33
13	511.63	28	500.51	43	489.86	58	479.67
14	510.87	29	499.78	44	489.17	59	479.00
15	510.12	30	499.06	45	488.48	60	478.34

RADII OF CURVES.

12°

'	RADIUS.	'	RADIUS.	'	RADIUS.	'	RADIUS.
0	478.34	15	468.61	30	459.28	45	450.31
1	477.68	16	467.98	31	458.67	46	449.72
2	477.02	17	467.35	32	458.06	47	449.14
3	476.36	18	466.72	33	457.45	48	448.56
4	475.71	19	466.09	34	456.85	49	447.97
5	475.05	20	465.46	35	456.25	50	447.39
6	474.40	21	464.83	36	455.65	51	446.82
7	473.75	22	464.21	37	455.05	52	446.24
8	473.10	23	463.59	38	454.45	53	445.67
9	472.46	24	462.97	39	453.85	54	445.09
10	471.81	25	462.35	40	453.26	55	444.52
11	471.17	26	461.73	41	452.66	56	443.95
12	470.53	27	461.11	42	452.07	57	443.38
13	469.89	28	460.50	43	451.48	58	442.81
14	469.25	29	459.89	44	450.89	59	442.25
15	468.61	30	459.28	45	450.31	60	441.68

RADII OF CURVES.

13°

'	RADIUS.	'	RADIUS.	'	RADIUS.	'	RADIUS.
0	441.68	15	433.39	30	425.40	45	417.70
1	441.12	16	432.84	31	424.87	46	417.20
2	440.56	17	432.30	32	424.35	47	416.69
3	440.00	18	431.76	33	423.83	48	416.19
4	439.44	19	431.23	34	423.32	49	415.69
5	438.88	20	430.69	35	422.80	50	415.19
6	438.33	21	430.15	36	422.28	51	414.70
7	437.77	22	429.62	37	421.77	52	414.20
8	437.22	23	429.09	38	421.26	53	413.71
9	436.67	24	428.56	39	420.74	54	413.21
10	436.12	25	428.03	40	420.23	55	412.72
11	435.57	26	427.50	41	419.72	56	412.23
12	435.02	27	426.97	42	419.22	57	411.74
13	434.47	28	426.45	43	418.71	58	411.25
14	433.93	29	425.92	44	418.20	59	410.76
15	433.39	30	425.40	45	417.70	60	410.26

RADII OF CURVES.

14°

′	RADIUS.	′	RADIUS.	′	RADIUS.	′	RADIUS.
0	410.28	15	403.11	30	396.20	45	389.52
1	409.79	16	402.65	31	395.75	46	389.08
2	409.31	17	402.18	32	395.30	47	388.65
3	408.82	18	401.71	33	394.85	48	388.21
4	408.34	19	401.25	34	394.40	49	387.78
5	407.86	20	400.78	35	393.95	50	387.34
6	407.38	21	400.32	36	393.50	51	386.91
7	406.90	22	399.86	37	393.05	52	386.48
8	406.42	23	399.40	38	392.61	53	386.05
9	405.95	24	398.94	39	392.16	54	385.62
10	405.47	25	398.48	40	391.72	55	385.19
11	405.00	26	398.02	41	391.28	56	384.77
12	404.53	27	397.56	42	390.84	57	384.34
13	404.05	28	397.11	43	390.40	58	383.91
14	403.58	29	396.65	44	389.96	59	383.49
15	403.11	30	396.20	45	389.52	60	383.06

RADII OF CURVES.

15°

'	RADIUS.	'	RADIUS.	'	RADIUS.	'	RADIUS.
0	383.06	15	376.82	30	370.78	45	364.93
1	382.64	16	376.41	31	370.38	46	364.55
2	382.22	17	376.00	32	369.99	47	364.16
3	381.80	18	375.60	33	369.60	48	363.78
4	381.38	19	375.19	34	369.20	49	363.40
5	380.96	20	374.79	35	368.81	50	363.02
6	380.54	21	374.38	36	368.42	51	362.64
7	380.13	22	373.98	37	368.03	52	362.26
8	379.71	23	373.57	38	367.64	53	361.89
9	379.29	24	373.17	39	367.25	54	361.51
10	378.88	25	372.77	40	366.86	55	361.13
11	378.47	26	372.37	41	366.47	56	360.76
12	378.05	27	371.97	42	366.09	57	360.38
13	377.64	28	371.57	43	365.69	58	360.01
14	377.23	29	371.18	44	365.31	59	359.64
15	376.82	30	370.78	45	364.93	60	359.27

RADII OF CURVES.

16°

'	RADIUS.	'	RADIUS.	'	RADIUS.	'	RADIUS.
0	359.27	15	353.77	30	348.45	45	343.29
1	358.89	16	353.41	31	348.10	46	342.95
2	358.52	17	353.05	32	347.75	47	342.61
3	358.15	18	352.70	33	347.40	48	342.27
4	357.78	19	352.34	34	347.06	49	341.93.
5	357.42	20	351.98	35	346.71	50	341.60
6	357.05	21	351.62	36	346.37	51	341.26
7	356.68	22	351.27	37	346.02	52	340.93
8	356.31	23	350.91	38	345.68	53	340.59
9	355.95	24	350.56	39	345.33	54	340.26
10	355.59	25	350.21	40	344.99	55	339.93
11	355.22	26	349.85	41	344.65	56	339.59
12	354.86	27	349.50	42	344.31	57	339.26
13	354.50	28	349.15	43	343.97	58	338.93
14	354.13	29	348.80	44	343.63	59	338.60
15	353.77	30	348.45	45	343.29	60	338.27

RADII OF CURVES.

17°

'	RADIUS.	'	RADIUS.	'	RADIUS.	'	RADIUS.
0	338.27	15	333.41	30	328.68	45	324.09
1	337.94	16	333.09	31	328.37	46	323.79
2	337.62	17	332.77	32	328.06	47	323.49
3	337.29	18	332.45	33	327.75	48	323.18
4	336.96	19	332.13	34	327.44	49	322.89
5	336.64	20	331.82	35	327.14	50	322.59
6	336.31	21	331.50	36	326.83	51	322.29
7	335.99	22	331.18	37	326.52	52	321.99
8	335.66	23	330.87	38	326.22	53	321.69
9	335.34	24	330.56	39	325.91	54	321.39
10	335.01	25	330.24	40	325.60	55	321.10
11	334.69	26	329.93	41	325.30	56	320.80
12	334.37	27	329.62	42	325.00	57	320.51
13	334.05	28	329.30	43	324.70	58	320.21
14	333.73	29	328.99	44	324.39	59	319.92
15	333.41	30	328.68	45	324.09	60	319.62

RADII OF CURVES.

18°

'	RADIUS.	'	RADIUS.	'	RADIUS.	'	RADIUS.
0	319.62	15	315.28	30	311.06	45	306.95
1	319.33	16	315.00	31	310.78	46	306.68
2	319.04	17	314.71	32	310.50	47	306.41
3	318.74	18	314.43	33	310.23	48	306.14
4	318.45	19	314.14	34	309.95	49	305.87
5	318.16	20	313.86	35	309.67	50	305.60
6	317.87	21	313.58	36	309.40	51	305.33
7	317.58	22	313.29	37	309.12	52	305.06
8	317.29	23	313.01	38	308.85	53	304.80
9	317.00	24	312.73	39	308.58	54	304.53
10	316.71	25	312.45	40	308.30	55	304.27
11	316.43	26	312.17	41	308.03	56	304.00
12	316.14	27	311.89	42	307.76	57	303.73
13	315.85	28	311.61	43	307.49	58	303.47
14	315.57	29	311.33	44	307.22	59	303.21
15	315.28	30	311.06	45	306.95	60	302.94

RADII OF CURVES.

18°

´	RADIUS.	´	RADIUS.	´	RADIUS.	´	RADIUS.
0	302.94	15	299.04	30	295.25	45	291.55
1	302.63	16	298.79	31	295.00	46	291.30
2	302.42	17	298.53	32	294.75	47	291.06
3	302.16	18	298.23	33	294.50	48	290.82
4	301.89	19	298.02	34	294.25	49	290.58
5	301.63	20	297.77	35	294.00	50	290.33
6	301.37	21	297.51	36	293.76	51	290.09
7	301.11	22	297.26	37	293.51	52	289.85
8	300.85	23	297.01	38	293.26	53	289.61
9	300.59	24	296.75	39	293.01	54	289.37
10	300.33	25	296.50	40	292.77	55	289.13
11	300.07	26	296.25	41	292.52	56	288.89
12	299.82	27	296.00	42	292.28	57	288.65
13	299.56	28	295.75	43	292.03	58	288.41
14	299.30	29	295.50	44	291.79	59	288.18
15	299.04	30	295.25	45	291.55	60	287.94

RADII OF CURVES.

20°

′	RADIUS.	′	RADIUS.	′	RADIUS.	′	RADIUS.
0	287.94	15	284.42	30	280.99	45	277.64
1	287.70	16	284.19	31	280.76	46	277.42
2	287.46	17	283.96	32	280.54	47	277.20
3	287.23	18	283.73	33	280.31	48	276.98
4	286.99	19	283.50	34	280.09	49	276.76
5	286.76	20	283.27	35	279.86	50	276.54
6	286.52	21	283.04	36	279.64	51	276.32
7	286.29	22	282.81	37	279.42	52	276.10
8	286.05	23	282.58	38	279.19	53	275.89
9	285.82	24	282.35	39	278.97	54	275.67
10	285.58	25	282.12	40	278.75	55	275.45
11	285.35	26	281.89	41	278.52	56	275.23
12	285.12	27	281.67	42	278.30	57	275.02
13	284.88	28	281.44	43	278.08	58	274.80
14	284.65	29	281.21	44	277.86	59	274.58
15	284.42	30	280.99	45	277.64	60	274.37

RADII OF CURVES.

21°

′	RADIUS.	′	RADIUS.	′	RADIUS.	′	RADIUS.
0	274.33	15	271.18	30	268.06	45	265.02
1	274.12	16	270.97	31	267.86	46	264.82
2	273.91	17	270.76	32	267.65	47	264.62
3	273.73	18	270.55	33	267.45	48	264.42
4	273.51	19	270.34	34	267.24	49	264.22
5	273.30	20	270.13	35	267.04	50	264.02
6	273.08	21	269.92	36	266.84	51	263.82
7	272.87	22	269.71	37	266.63	52	263.62
8	272.66	23	269.51	38	266.43	53	263.42
9	272.45	24	269.30	39	266.23	54	263.22
10	272.23	25	269.09	40	266.02	55	263.03
11	272.02	26	268.89	41	265.82	56	262.83
12	271.81	27	268.68	42	265.62	57	262.63
13	271.60	28	268.47	43	265.42	58	262.41
14	271.39	29	268.27	44	265.22	59	262.24
15	271.18	30	268.06	45	265.02	60	262.04

RADII OF CURVES.

22°

'	RADIUS.	'	RADIUS.	'	RADIUS.	'	RADIUS.
0	262.04	15	259.13	30	256.29	45	253.51
1	261.85	16	258.94	31	256.10	46	253.33
2	261.65	17	258.75	32	255.92	47	253.14
3	261.45	18	258.56	33	255.73	48	252.96
4	261.26	19	258.37	34	255.54	49	252.78
5	261.06	20	258.18	35	255.36	50	252.60
6	260.87	21	257.99	36	255.17	51	252.42
7	260.68	22	257.80	37	254.99	52	252.24
8	260.48	23	257.61	38	254.80	53	252.05
9	260.29	24	257.42	39	254.62	54	251.87
10	260.10	25	257.23	40	254.43	55	251.69
11	259.90	26	257.04	41	254.25	56	251.51
12	259.71	27	256.85	42	254.06	57	251.33
13	259.52	28	256.67	43	254.88	58	251.15
14	259.33	29	256.48	44	253.70	59	250.97
15	259.13	30	256.29	45	253.51	60	250.79

RADII OF CURVES.

25°

′	RADIUS.	′	RADIUS.	′	RADIUS.	′	RADIUS.
0	250.79	15	248.13	30	245.53	45	242.98
1	250.61	16	247.96	31	245.36	46	242.81
2	250.43	17	247.78	32	245.19	47	242.64
3	250.26	18	247.61	33	245.02	48	242.48
4	250.08	19	247.43	34	244.84	49	242.31
5	249.90	20	247.26	35	244.67	50	242.14
6	249.72	21	247.08	36	244.50	51	241.98
7	249.54	22	246.91	37	244.33	52	241.81
8	249.37	23	246.74	38	244.16	53	241.64
9	249.19	24	246.56	39	243.99	54	241.48
10	249.01	25	246.39	40	243.83	55	241.31
11	249.84	26	246.22	41	243.66	56	241.15
12	248.66	27	246.04	42	243.49	57	240.98
13	248.48	28	245.87	43	243.32	58	240.82
14	248.31	29	245.70	44	243.15	59	240.65
15	248.13	30	245.53	45	242.98	60	240.49

CATALOGUE

OF THE

SCIENTIFIC PUBLICATIONS

OF

D. VAN NOSTRAND COMPANY,

23 MURRAY STREET AND 27 WARREN STREET, N. Y.

ADAMS (J. W.) Sewers and Drains for Populous Districts. 8vo, cloth.................................$2 50

ALEXANDER (J. H.) Universal Dictionary of Weights and Measures. 8vo, cloth 3 50

—— (S. A.) Broke Down: What Should I Do? A Ready Reference and Key to Locomotive Engineers and Firemen, Round-house Machinists, Conductors, Train Hands and Inspectors. With 5 folding plates. 12mo, cloth.. 1 50

ATKINSON (PHILIP). The Elements of Electric Lighting, including Electric Generation, Measurements, Storage, and Distribution. Seventh edition. Illustrated. 12mo, cloth...................................... 1 50

—— The Elements of Dynamic Electricity and Magnetism. 120 illustrations. 12mo, cloth.................... 2 00

—— Elements of Static Electricity, with full description of the Holtz and Topler Machines, and their mode of operating. Illustrated. 12mo, cloth................. 1 50

—— The Electric Transformation of Power and its Application by the Electric Motor, including Electric Railway Construction. Illustrated. 12mo, cloth....... 2 00

AUCHINCLOSS (W. S.) Link and Valve Motions Simplified. Illustrated with 37 woodcuts and 21 lithographic plates, together with a Travel Scale, and numerous useful tables. Eleventh edition. 8vo, cloth.. 3 00

BACON (F. W.) A Treatise on the Richards Steam-Engine Indicator, with directions for its use. By Charles T. Porter. Revised. Illustrated. 12mo, cloth....... 1 00

BADT (F. B.) Dynamo Tender's Hand-book. With 70 illustrations. Second edition. 18mo, cloth........... 1 00

—— Bell-hangers' Hand-book. With 97 illustrations. 18mo, cloth........ .... 1 00

—— Incandescent Wiring Hand-book. With 35 illustrations and five tables. 18mo, cloth................. 1 00

—— Electric Transmission Hand-book. With 22 illustrations and 27 tables. 18mo, cloth................. 1 00

BALE (M. P.) Pumps and Pumping. A Hand-book for Pump Users. 12mo, cloth....................... 1 00

BARBA (J.) The Use of Steel for Constructive Purposes. Method of Working, Applying, and Testing Plates and Bars. With a Preface by A. L. Holley, C.E. 12mo, cloth... .... 1 50

BARNARD (F. A. P.) Report on Machinery and Processes of the Industrial Arts and Apparatus of the Exact Sciences at the Paris Universal Exposition, 1867. 152 illustrations and 8 folding plates. 8vo, cloth.... .... 5 00

BEAUMONT (ROBERT). Color in Woven Design. With 32 colored Plates and numerous original illustrations. Large 12mo............................... 7 50

BEILSTEIN (F.) An Introduction to Qualitative Chemical Analysis. Translated by I. J. Osbun. 12mo, cloth.... ....

BECKWITH (ARTHUR). Pottery. Observations on the Materials and Manufacture of Terra-Cotta, Stoneware, Fire-brick, Porcelain, Earthenware, Brick, Majolica, and Encaustic Tiles. 8vo, paper....... ... 60

BERNTHSEN (A.) A Text-book of Organic Chemistry. Translated by George McGowan, Ph.D. 544 pages. Illustrated. 12mo, cloth.............. .... 2 50

BIGGS (C. H. W.) First Principles of Electrical Engineering. 12mo, cloth. Illustrated................... 1 00

BLAKE (W. P.) Report upon the Precious Metals. 8vo, cloth 2 00

—— Ceramic Art. A Report on Pottery, Porcelain, Tiles, Terra-Cotta, and Brick. 8vo, cloth 2 00

BLAKESLEY (T. H.) Alternating Currents of Electricity. For the use of Students and Engineers. 12mo, cloth. 1 50

BLYTH (A. WYNTER, M.R.C.S., F.C.S.) Foods : their Compositions and Analysis. Crown 8vo, cloth.. .... 6 00

—— Poisons: their Effects and Detection. Crown 8vo, cloth...... .... 6 00

BODMER (G. R.) Hydraulic Motors ; Turbines and Pressure Engines, for the use of Engineers, Manufacturers, and Students. With numerous illustrations. 12mo, cloth...... .... 5 00

BOTTONE (S. R.) Electrical Instrument Making for Amateurs. With 48 illustrations. 12mo, cloth 50

—— Electric Bells, and all about them. Illustrated. 12mo, cloth 50

—— The Dynamo: How Made and How Used. 12mo, cloth.... .... 1 00

—— Electro Motors: How Made and How Used. 12mo. cloth...... ... .•...................... 50

BONNEY (G. E.) The Electro-Platers' Hand-book. 60 Illustrations. 12mo, cloth.... .... 1 20

BOW (R. H.) A Treatise on Bracing. With its application to Bridges and other Structures of Wood or Iron. 156 illustrations. 8vo, cloth...................... .... 1 50

BOWSER (Prof. E. A.) An Elementary Treatise on Analytic Geometry. Embracing plain Geometry, and an Introduction to Geometry of three Dimensions. 12mo, cloth. Thirteenth edition.... .... 1 75

—— An Elementary Treatise on the Differential and Integral Calculus. With numerous examples. 12mo, cloth. Twelfth edition 2 25

—— An Elementary Treatise on Analytic Mechanics. With numerous examples. 12mo, cloth. Fifth edition. 3 00

BOWSER (Prof. E. A.) An Elementary Treatise on Hydro-mechanics. With numerous examples. 12mo, cloth. Third edition........................ .... 2 5c

BOWIE (AUG. J., Jun., M. E.) A Practical Treatise on Hydraulic Mining in California. With Description of the Use and Construction of Ditches, Flumes, Wrought-iron Pipes, and Dams; Flow of Water on Heavy Grades, and its Applicability, under High Pressure, to Mining. Third edition. Small quarto, cloth. Illustrated........ .... 5 00

BURGH (N. P.) Modern Marine Engineering, applied to Paddle and Screw Propulsion. Consisting of 36 colored plates, 259 practical woodcut illustrations, and 403 pages of descriptive matter. Thick 4to vol., half morocco................................. ....10 00

BURT (W A.) Key to the Solar Compass, and Surveyor's Companion. Comprising all the rules necessary for use in the field. Pocket-book form, tuck 2 50

CALDWELL (Prof. GEO. C., and BRENEMAN (Prof. A. A.) Manual of Introductory Chemical Practice. 8vo, cloth. Illustrated........ .. .. 1 50

CAMPIN (FRANCIS). On the Construction of Iron Roofs. A Theoretical and Practical Treatise, with wood-cuts and plates of Roofs recently executed. 8vo, cloth........... .... 2 00

CLEEMAN (THOS. M.) The Railroad Engineer's Practice. Being a Short but Complete Description of the Duties of the Young Engineer in the Preliminary and Location Surveys and in Construction. Fourth edition, revised and enlarged. Illustrated, 12mo, cloth....... 2 00

CLARK (D. KINNEAR, C.E.) A Manual of Rules, Tables and Data for Mechanical Engineers. Illustrated with numerous diagrams. 1012 pages. 8vo, cloth 5 00
Half morocco.. 7 50

——— Fuel; its Combustion and Economy, consisting of abridgments of Treatise on the Combustion of Coal. By C. W. Williams; and the Economy of Fuel, by T. S. Prideaux. With extensive additions in recent practice in the Combustion and Economy of Fuel, Coal, Coke, Wood, Peat, Petroleum, etc. 12mo, cloth. 1 50

DORR (B. F) The Surveyor's Guide and Pocket Table Book. 18mo, morocco flaps. Second edition 2 00

DUBOIS (A. J.) The New Method of Graphic Statics. With 60 illustrations. 8vo, cloth 1 50

EDDY (Prof. H. T.) Researches in Graphical Statics. Embracing New Constructions in Graphical Statics, a New General Method in Graphical Statics, and the Theory of Internal Stress in Graphical Statics. 8vo, cloth......... .... 1 50

—— Maximum Stresses under Concentrated Loads. Treated graphically. Illustrated. 8vo, cloth... .... 1 50

EISSLER (M.) The Metallurgy of Gold; a Practical Treatise on the Metallurgical Treatment of Gold-Bearing Ores. 187 illustrations. 12mo, cl............ 5 00

—— The Metallurgy of Silver; a Practical Treatise on the Amalgamation, Roasting, and Lixiviation of Silver Ores. 124 illustrations. 12mo, cloth 4 00

—— The Metallurgy of Argentiferous Lead ; a Practical Treatise on the Smelting of Silver-Lead Ores and the refining of Lead Bullion. With 183 illustrations. 3vo, cloth 5 00

ELIOT Prof. C. W.) and STORER (Prof F. H.) A Compendious Manual of Qualitative Chemical Analys s. Revised with the co-operation of the authors, by Prof. William R. Nichols. Illustrated. 17th edition. Newly revised by Prof. W. B. Lindsay. 12mo, cloth 1 50

EVERETT (J. D.) Elementary Text-book of Physics. Illustrated. 12mo, cloth 1 40

FANNING (J. T.) A Practical Treatise on Hydraulic and Water-supply Engineering. Relating to the Hydrology. Hydrodynamics, and Practical Construction of Water-works in North America. Illustrated. 8vo, cloth......... .... 5 00

FISKE (Lieut. BRADLEY A., U. S. N.) Electricity in Theory and Practice ; or, The Elements of Electrical Engineering. 8vo, cloth............... .... 2 50

FLEMING (Prof. A. J.) The Alternate Current Transformer in Theory and Practice. Vol. I.—The Induction of Electric Currents. Illustrated. 8vo, cloth.... 3 00

FOLEY (NELSON), and THOS. PRAY, Jr. The Mechanical Engineers' Reference-book for Machine and Boiler Construction, in two parts. Part I—General Engineering Data. Part 2—Boiler Construction. With fifty-one plates and numerous illustrations, specially drawn for this work. Folio, half mor........25 00

FORNEY (MATTHIAS N.) Catechism of the Locomotive. Revised and enlarged. 8vo, cloth. 3 50

FOSTER (Gen. J. G., U. S. A.) Submarine Blasting in Boston Harbor, Massachusetts. Removal of Tower and Corwin Rocks. Illustrated with 7 plates. 4to, cloth... 3 50

FRANCIS (Jas. B., C.E.) Lowell Hydraulic Experiments. Being a selection from experiments on Hydraulic Motors, on the Flow of Water over Weirs, in open Canals of uniform rectangular section, and through submerged Orifices and diverging Tubes. Made at Lowell, Mass. Illustrated. 4to, cloth......15 00

GERBER (NICHOLAS). Chemical and Physical Analysis of Milk, Condensed Milk, and Infant's Milk-Food. 8vo, cloth................................. ... 1 25

GILLMORE (Gen. Q. A.) Treatise on Limes, Hydraulic Cements, and Mortars. With numerous illustrations. 8vo, cloth.... .... 4 00

—— Practical Treatise on the Construction of Roads, Streets, and Pavements. With 70 illustrations. 12mo, cloth............................... ...... 2 00

—— Report on Strength of the Building-Stones in the United States, etc. Illustrated. 8vo, cloth.... ...... 1 00

GOODEVE (T. M.) A Text-book on the Steam-Engine With a Supplement on Gas-Engines. 143 illustrations. 12mo, cloth. 2 00

GORE (G., F.R.S.) The Art of Electrolytic Separation of Metals, etc. (Theoretical and Practical.) Illustrated. 8vo, cloth 3 50

GORDON (J. E. H.) School Electricity. Illustrations. 12mo, cloth.......... .... 2 00

GRIMSHAW (ROBERT, M.E.) The Steam Boiler. Catechism. A Practical Book for Steam Engineers, Firemen and Owners and Makers of Boilers of any kind. Illustrated. Thick 18mo, cloth...............

GRIFFITHS (A. B., Ph.D.) A Treatise on Manures, or the Philosophy of Manuring. A Practical Hand-book for the Agriculturist, Manufacturer, and Student. 12mo, cloth.. 3 00

GRUNER (M. L.) The Manufacture of Steel. Translated from the French, by Lenox Smith; with an appendix on the Bessemer process in the United States, by the translator. Illustrated. 8vo, cloth...... ...... 3 50

GURDEN (RICHARD LLOYD). Traverse Tables: computed to 4 places Decimals for every ° of angle up to 100 of Distance. For the use of Surveyors and Engineers. New edition. Folio, half mor........... 7 50

HALSEY (F. A.) Slide-valve Gears, an Explanation of the Action and Construction of Plain and Cut-off Slide Valves. Illustrated. 12mo, cloth. Second edition.. 1 50

HAMILTON (W. G.) Useful Information for Railway Men. Tenth edition, revised and enlarged. 562 pages, pocket form. Morocco, gilt 2 00

HARRISON (W. B.) The Mechanics' Tool Book. With Practical Rules and Suggestions for use of Machinists, Iron-Workers, and others. Illustrated with 44 engravings. 12mo, cloth.............. .... 1 50

HASKINS (C. H.) The Galvanometer and its Uses. A Manual for Electricians and Students. 12mo, cloth.. 1 50

HAWKINS (C. C.) and WALLIS (F.) The Dynamo, Its Theory, Design and Manufacture. 8vo, cloth, 190 ills. 3 00

HEAP (Major D. P., U. S. A.) Electrical Appliances of the Present Day. Report of the Paris Electrical Exposition of 1881. 250 illustrations. 8vo, cloth. 2 00

HOUSTON (E. J.) Dictionary of Electrical Words, Terms and Phrases. Third edition, revised and enlarged. 8vo, cloth................................. 5 00

HERRMANN (GUSTAV). The Graphical Statics of
Mechanism. A Guide for the Use of Machinists,
Architects, and Engineers; and also a Text-book for
Technical Schools. Translated and annotated by
A. P. Smith, M.E. 12mo, cloth, 7 folding plates..... 2 00

HEWSON (WM.) Principles and Practice of Embanking
Lands from River Floods, as applied to the Levees of
the Mississippi. 8vo, cloth... .. .. 2 00

HENRICI (OLAUS). Skeleton Structures, Applied to
the Building of Steel and Iron Bridges. Illustrated.. 1 50

HOBBS (W. R. P.) The Arithmetic of Electrical Measurements, with numerous examples. 12mo, cloth.... 50

HOLLEY (ALEXANDER L.) Railway Practice. American and European Railway practice in the Economi-
cal Generation of Steam, including the Materials and
Construction of Coal-burning Boilers, Combustion,
the Variable Blast, Vaporization, Circulation, Super-
heating, Supplying and Heating Feed-water, etc.,
and the Adaptation of Wood and Coke-burning
Engines to Coal-burning; and in Permanent Way,
including Road-bed, Sleepers. Rails, Joint Fastenings,
Street Railways, etc. With 77 lithographed plates.
Folio, cloth......... ....12 00

HOLMES (A. BROMLEY). The Electric Light Popularly Explained. Fifth edition. Illustrated. 12mo,
paper. 40

HOWARD (C. R.) Earthwork Mensuration on the
Basis of the Prismoidal Formulæ. Containing Sim-
ple and Labor-saving Method of obtaining Prismoidal
Contents directly from End Areas. Illustrated by
Examples and accompanied by Plain Rules for Practi-
cal Uses. Illustrated. 8vo, cloth. 1 50

HUMBER (WILLIAM, C. E.) A Handy Book for the
Calculation of Strains in Girders, and Similar Structures, and their Strength; Consisting of Formulæ and
Corresponding Diagrams, with numerous details for
practical application, etc. Fourth edition. 12mo,
cloth.......... .... 2 50

HUTTON (W. S.) Steam-Boiler Construction. A Practical Hand-book for Engineers, Boiler Makers, and
Steam Users. With upwards of 300 illustrations.
8vo, cloth 7 00

ISHERWOOD (B. F.) Engineering Precedents for Steam Machinery. Arranged in the most practical and useful manner for Engineers. With illustrations. 2 vols. in 1. 8vo, cloth....... .... 2 50

JAMIESON (ANDREW, C.E.) A Text-book on Steam and Steam-Engines. Illustrated. 12mo, cloth....... 3 00

JANNETTAZ (EDWARD). A Guide to the Determination of Rocks; being an Introduction to Lithology. Translated from the French by Professor G. W. Plympton. 12mo, cloth............................ 1 50

JONES (H. CHAPMAN). Text-book of Experimental Organic Chemistry for Students. 18mo, cloth........ 1 00

JOYNSON (F. H.) The Metals used in Construction. Iron, Steel, Bessemer Metal, etc. Illustrated. 12mo, cloth................. 75

—— Designing and Construction of Machine Gearing. Illustrated. 8vo, cloth.............................. 2 00

KANSAS CITY BRIDGE (THE). With an Account of the Regimen of the Missouri River and a Description of the Methods used for Founding in that River. By O. Chanute, Chief Engineer, and George Morrison, Assistant Engineer. Illustrated with 5 lithographic views and 12 plates of plans. 4to, cloth.............. 6 00

KAPP (GISBERT, C.E.) Electric Transmission of Energy and its Transformation, Subdivision, and Distribution. A Practical Hand-book. 12mo, cloth..... 3 00

KEMPE (H. R.) The Electrical Engineer's Pocket Book of Modern Rules, Formulæ, Tables, and Data. Illustrated. 32mo, mor. gilt......................... 1 75

KENNELLEY (A. E.) Theoretical Elements of Electro-Dynamic Machinery. Vol. I. Illustrated. 8vo, cloth. 1 50

KING (W. H.) Lessons and Practical Notes on Steam. The Steam-Engine, Propellers, etc., for Young Marine Engineers, Students, and others. Revised by Chief Engineer J. W. King, United States Navy. 8vo, cloth............................. .... 2 00

KIRKALDY (WM. G.) Illustrations of David Kirkaldy's System of Mechanical Testing, as Originated and Carried On by him during a Quarter of a Century.

Comprising a Large Selection of Tabulated Results, showing the Strength and other Properties of Materials used in Construction, with Explanatory Text and Historical Sketch. Numerous engravings and 25 lithographed plates. 4to, cloth.................. ....25 00

KIRKWOOD (JAS. P.) Report on the Filtration of River Waters for the supply of Cities, as practised in Europe. Illustrated by 30 double-plate engravings. 4to, cloth.............................. ....15 00

LARRABEE (C. S.) Cipher and Secret Letter and Telegraphic Code, with Hog's Improvements. 18mo, cloth........ 60

LARDEN (W., M. A.) A School Course on Heat. 12mo, half leather........ .... 2 00

LEITZE (ERNST). Modern Heliographic Processes. A Manual of Instruction in the Art of Reproducing Drawings, Engravings, etc., by the action of Light. With 32 wood-cuts and ten specimens of Heliograms. 8vo, cloth. Second edition............ .... 3 00

LOCKWOOD (THOS. D.) Electricity, Magnetism, and (Electro-Telegraphy. A Practical Guide for Students, Operators, and Inspectors. 8vo, cloth. Third edition.................................... .... 2 50

LODGE (OLIVER J.) Elementary Mechanics, including Hydrostatics and Pneumatics. Revised edition. 12mo, cloth.............................. .... 1 20

LOCKE (ALFRED G. and CHARLES G.) A Practical Treatise on the Manufacture of Sulphuric Acid. With 77 Constructive Plates drawn to Scale Measurements, and other Illustrations. Royal 8vo, cloth 15 00

LOVELL (D. H.) Practical Switch Work. A Handbook for Track Foremen. Illustrated. 12mo, cloth.. 1 50

LUNGE (GEO.) A Theoretical and Practical Treatise on the Manufacture of Sulphuric Acid and Alkali with the Collateral Branches. Vol. I. Sulphuric Acid. Second edition, revised and enlarged. 342 Illustrations. 8vo., cloth................................. ....15 00

—— and HUNTER F.) The Alkali Maker's Pocket-Book. Tables and Analytical Methods for Manufacturers of Sulphuric Acid, Nitric Acid, Soda, Potash and Ammonia. Second edition. 12mo, cloth 3 00

MACCORD (Prof. C. W.) A Practical Treatise on the Slide-Valve by Eccentrics, examining by methods the action of the Eccentric upon the Slide-Valve, and explaining the practical processes of laying out the movements, adapting the Valve for its various duties in the Steam-Engine. Illustrated. 4to, cloth.... .... 2 50

MAYER (Prof. A. M.) Lecture Notes on Physics. 8vo. cloth.. 2 00

McCULLOCH (Prof. R. S.) Elementary Treatise on the Mechanical Theory of Heat, and its application to Air and Steam Engines. 8vo, cloth.............. ...... 3 50

MERRILL (Col. WM. E., U. S. A.) Iron Truss Bridges for Railroads. The method of calculating strains in Trusses, with a careful comparison of the most prominent Trusses, in reference to economy in combination, etc. Illustrated. 4to, cloth............................ 5 00

METAL TURNING. By a Foreman Pattern Maker. Illustrated with 81 engravings. 12mo, cloth.......... 1 50

MINIFIE (WM.) Mechanical Drawing. A Text-book of Geometrical Drawing for the use of Mechanics and Schools. in which the Definitions and Rules of Geometry are familiarly explained ; the Practical Problems are arranged from the most simple to the more complex. and in their description technicalities are avoided as much as possible. With illustrations for Drawing Plans, Sections, and Elevations of Railways and Machinery ; an Introduction to Isometrical Drawing, and an Essay on Linear Perspective and Shadows. Illustrated with over 200 diagrams engraved on steel. With an appendix on the Theory and Application of Colors. 8vo, cloth. 4 00

—— Geometrical Drawing. Abridged from the octavo edition, for the use of schools. Illustrated with 48 steel plates Ninth edition. 12mo, cloth 2 00

MODERN METEOROLOGY. A Series of Six Lectures, delivered under the auspices of the Meteorological Society in 1878. Illustrated. 12mo, cloth........... 1 50

MOONEY (WM.) The American Gas Engineers' and Superintendents' Hand-book, consisting of Rules, Reference Tables, and original matter pertaining to the Manufacture, Manipulation, and Distribution of Illuminating Gas. Illustrated. 12mo, morocco 3 00

MOTT (H. A., Jun.) A Practical Treatise on Chemistry (Qualitative and Quantitative Analysis), Stoichiometry, Blow-pipe Analysis, Mineralogy, Assaying, Pharmaceutical Preparations. Human Secretions, Specific Gravities, Weights and Measures, etc. New Edition, 1883. 650 pages. 8vo, cloth... .... 4 00

MULLIN (JOSEPH P., M.E.) Modern Moulding and Pattern-making. A Practical Treatise upon Pattern-Shop and Foundry Work: embracing the Moulding of Pulleys, Spur Gears, Worm Gears, Balance-wheels, Stationary Engine and Locomotive Cylinders, Globe Valves, Tool Work, Mining Machinery, Screw Propellers, Pattern-shop Machinery, and the latest improvements in English and American Cupolas; together with a large collection of original and carefully selected Rules and Tables for every-day use in the Drawing Office, Pattern-shop, and Foundry. 12mo, cloth, illustrated.............................. 2 50

MUNRO (JOHN, C.E.) and JAMIESON (ANDREW, C.E.) A Pocket-book of Electrical Rules and Tables for the use of Electricians and Engineers. Seventh edition, revised and enlarged. With numerous diagrams. Pocket size. Leather............... 2 50

MURPHY (J. G., M.E.) Practical Mining. A Field Manual for Mining Engineers. With Hints for Investors in Mining Properties. 16mo, morocco tucks.. 1 50

NAQUET (A.) Legal Chemistry. A Guide to the Detection of Poisons, Falsification of Writings, Adulteration of Alimentary and Pharmaceutical Substances, Analysis of Ashes, and examination of Hair, Coins, Arms, and Stains, as applied to Chemical Jurisprudence, for the use of Chemists, Physicians, Lawyers, Pharmacists and Experts. Translated, with additions, including a list of books and memoirs on Toxicology, etc., from the French, by J. P. Battershall, Ph.D., with a preface by C. F. Chandler, Ph.D., M.D., LL.D. 12mo, cloth..................... .... 2 00

NEWALL (J. W.) Plain Practical Directions for Drawing, Sizing and Cutting Bevel-Gears, showing how the Teeth may be cut in a plain Milling Machine or Gear Cutter so as to give them a correct shape, from end to end; and showing how to get out all particulars for the Workshop without making any Drawings. Including a full set of Tables of Reference. Folding Plates, 8vo., cloth. 3 00

NEWLANDS (JAMES). The Carpenter's and Joiners' Assistant : being a Comprehensive Treatise on the Selection, Preparation and Strength of Materials, and the Mechanical Principles of Framing, with their application in Carpentry, Joinery, and Hand-Railing ; also, a Complete Treatise on Sines ; and an illustrated Glossary of Terms used in Architecture and Building. Illustrated. Folio, half mor.........................15 00

NIBLETT (J. T.) Secondary Batteries. Illustrated. 12mo, cloth 1 50

NIPHER (FRANCIS E., A M) Theory of Magnetic Measurements, with an appendix on the Method of Least Squares. 12mo, cloth...................... 1 00

NOAD (HENRY M.) The Students' Text-book of Electricity. A new edition, carefully revised. With an Introduction and additional chapters by W. H. Preece. With 471 illustrations. 12mo, cloth. 4 00

NUGENT (E.) Treatise on Optics ; or, Light and Sight theoretically and practically treated, with the application to Fine Art and Industrial Pursuits. With 103 illustrations. 12mo. cloth. 1 50

PAGE (DAVID). The Earth's Crust, a Handy Outline of Geology. 16mo, cloth........................... 75

PARSONS (Jr., W. B., C.E) Track, a Complete Manual of Maintenance of Way, according to the Latest and Best Practice on Leading American Railroads. Illustrated. 8vo, cloth 2 00

PEIRCE (B.) System of Analytic Mechanics. 4to, cloth 10 00

PHILLIPS (JOSHUA). Engineering Chemistry. A Practical Treatise for the use of Analytical Chemists, Engineers, Iron Masters, Iron Founders, students and others. Comprising methods of Analysis and Valuation of the principal materials used in Engineering works, with numerous Analyses, Examples and Suggestions. 314 Illustrations. 8vo, cloth.... .... 4 00

PLANE TABLE (THE). Its Uses in Topographical Surveying. Illustrated. 8vo, cloth.................. 2 00

PLATTNER. Manual of Qualitative and Quantitative Analysis with the Blow-pipe. From the last German edition, revised and enlarged, by Prof. Th. Richter, of the Royal Saxon Mining Academy. Translated by

Prof. H. B. Cornwall, assisted by John H. Caswell. Illustrated with 87 wood-cuts and one lithographic plate. Fourth edition, revised. 560 pages. 8vo, cloth...... .... 5 00

PLANTE (GASTON). The Storage of Electrical Energy, and Researches in the Effects created by Currents, combining Quantity with High Tension. Translated from the French by Paul B. Elwell. 89 illustrations. 8vo............................ .... 4 00

PLYMPTON (Prof. GEO. W.) The Blow-pipe. A Guide to its use in the Determination of Salts and Minerals. Compiled from various sources. 12mo, cloth......... .... 1 50

POCKET LOGARITHMS to Four Places of Decimals, including Logarithms of Numbers and Logarithmic Sines and Tangents to Single Minutes. To which is added a Table of Natural Sines, Tangents and Co-Tangents. 16mo, boards..................... 50

POPE (F. L.) Modern Practice of the Electric Telegraph. A Technical Hand-book for Electricians, Managers and Operators. New edition, rewritten and enlarged, and fully illustrated. 8vo, cloth...... 1 50

PRAY (Jr., THOMAS). Twenty Years with the Indicator: being a Practical Text-book for the Engineer or the Student. Illustrated. 8vo, cloth.............. 2 50

PRACTICAL IRON-FOUNDING. By the author of "Pattern Making," etc., etc. Illustrated with over one hundred engravings. 12mo, cloth.... .... 1 50

PREECE (W H.) and STUBBS (A. J.) Manual of Telephony. Illustrations and Plates. 12mo, cloth........ 4 50

PRESCOTT (Prof. A. B.) Organic Analysis. A Manual of the Descriptive and Analytical Chemistry of certain Carbon Compounds in Common Use; a Guide in the Qualitative and Quantitative Analysis of Organic Materials in Commercial and Pharmaceutical Assays, in the estimation of Impurities under Authorized Standards, and in Forensic Examinations for Poisons, with Directions for Elementary Organic Analysis. 8vo, cloth........ .... 5 00

—— Outlines of Proximate Organic Analysis, for the Identification, Separation, and Quantitative Determination of the more commonly occurring Organic Compounds. 12mo, cloth 1 75

PRESCOTT (Prof. A. B.) First Book in Qualitative
Chemistry. Fifth edition. 12mo, cloth.......... .. 1 50
—— and OTIS COE JOHNSON. Qualitative Chemical
Analysis. A Guide in the Practical Study of Chem-
istry and in the work of Analysis. Revised edition
With Descriptive Chemistry extended throughout.... 3 50
PRITCHARD (O. G.) The Manufacture of Electric
Light Carbons. Illustrated. 8vo, paper........ 60
PULSIFER (W. H.) Notes for a History of Lead. 8vo,
cloth, gilt tops 4 00
PYNCHON Prof. T. R.) Introduction to Chemical
Physics, designed for the use of Academies, Colleges,
and High Schools. 269 illustrations on wood. Crown
8vo, cloth 3 00
RANDALL (J. E.) A Practical Treatise on the Incan-
descent Lamp. Illustrated. 16mo, cloth............ 50
—— (P. M.) Quartz Operator's Hand-book. New edi-
tion, revised and enlarged, fully illustrated. 12mo,
cloth.... ... . 2 00
RAFTER (GEO. W.) Sewage Disposal in the United
States. Illustrated. 8vo, cloth.................... 6 00
RANKINE (W. J. MACQUORN, C.E., LL.D., F.R.S.)
Applied Mechanics. Comprising the Principles of
Statics and Cinematics, and Theory of Structures,
Mechanism, and Machines. With numerous dia-
grams. Thoroughly revised by W. J. Millar. Crown
8vo, cloth................................... 5 00
—— Civil Engineering. Comprising Engineering Sur-
veys, Earthwork, Foundations, Masonry, Carpentry,
Metal-work, Roads, Railways, Canals, Rivers, Water-
Works, Harbors, etc. With numerous tables and
illustrations. Thoroughly revised by W. J. Millar.
Crown 8vo, cloth... ... . 6 50
—— Machinery and Millwork. Comprising the Geom-
try, Motions, Work, Strength, Construction, and
Objects of Machines, etc. Illustrated with nearly 300
woodcuts. Thoroughly revised by W. J. Miller.
Crown 8vo, cloth....... .... 5 00
—— The Steam-Engine and Other Prime Movers.
With diagram of the Mechanical Properties of Steam,
folding plates, numerous tables and illustrations.
Thoroughly revised by W. J. Millar. Crown 8vo,
cloth... ... 5 00

RANKINE (W. J. MACQUORN. C.E., LL.D., F.R.S.
Useful Rules and Tables for Engineers and Others.
With Appendix. tables. tests, and formulæ for the use
of Electrical Engineers Comprising Submarine
Electrical Engineering, Electric Lighting, and Trans-
mission of Power. By Andrew Jamieson, C.E,
F.R.S.E. Thoroughly revised by W. J. Mill.
Crown 8vo, cloth 4 00

——— A Mechanical Text-book. By Prof. Macquorn
Rankine and E. F. Bamber, C.E. With numerous
illustrations. Crown, 8vo, cloth. 3 50

REED'S ENGINEERS' HAND-BOOK, to the Local
Marine Board Examinations for Certificates of Com-
petency as First and Second Class Engineers. By
W. H. Thorn. Illustrated. 8vo, cloth. 4 50

RICE (Prof. J. M.) and JOHNSON (Prof. W. W.) On a
New Method of obtaining the Differential of Functions, with especial reference to the Newtonian Con-
ception of Rates or Velocities. 12mo, paper......... 50

RIPPER (WILLIAM). A Course of Instruction in Machine Drawing and Design for Technical Schools and
Engineer Students. With 52 plates and numerous
explanatory engravings. Folio, cloth.............. 7 50

ROEBLING (J. A.) Long and Short Span Railway
Bridges. Illustrated with large copperplate engravings of plans and views. Imperial folio, cloth..... . 25 00

ROGERS (Prof. H. D.) The Geology of Pennsylvania.
A Government Survey, with a General View of the
Geology of the United States. essays on the Coal
Formation and its Fossils. and a description of the
Coal Fields of North America and Great Britain.
Illustrated with plates and engravings in the text. 3
vols. 4to, cloth, with portfolio of maps.............15 00

ROSE (JOSHUA. M.E) The Pattern-makers' Assistant.
Embracing Lathe Work. Branch Work, Core Work,
Sweep Work. and Practical Gear Constructions, the
Preparation and Use of Tools. together with a large
collection of useful and valuable Tables. Sixth
edition. Illustrated with 250 engravings. 8vo, cloth. 2 50

——— Key to Engines and Engine-Running. A Practical Treatise upon the Management of Steam Engines
and Boilers. for the Use of Those who Desire to Pass

an Examination to Take Charge of an Engine or Boiler. With numerous illustrations, and Instructions upon Engineers' Calculations, Indicators, Diagrams, Engine Adjustments, and other Valuable Information necessary for Engineers and Firemen. 12mo, cloth. 3 00

SABINE (ROBERT). History and Progress of the Electric Telegraph. With descriptions of some of the apparatus. 12mo, cloth................... ...... 1 25

SAELTZER (ALEX.) Treatise on Acoustics in connection with Ventilation. 12mo, cloth....... .... 1 00

SALOMONS (Sir DAVID, M. A.) Electric Light Installations. Vol. I. The management of Accumulators. Seventh edition, revised and enlarged, with numerous illustrations. 12mo, cloth........................... 1.50

SAUNNIER (CLAUDIUS). Watchmaker's Hand-book. A Workshop Companion for those engaged in Watchmaking and allied Mechanical Arts. Translated by J. Tripplin and E. Rigg. 12mo, cloth............... 3 50

SEATON (A. E.) A Manual of Marine Engineering. Comprising the Designing, Construction, and Working of Marine Machinery. With numerous tables and illustrations. 10th edition. 8vo, cloth............... 5 00

SCHUMANN (F.) A Manual of Heating and Ventilation in its Practical Application. for the use of Engineers and Architects. Embracing a series of Tables and Formulæ for dimensions of heating, flow and return pipes for steam and hot-water boilers, flues, etc. 12mo, illustrated, full roan................... ...... 1 50

—— Formulas and Tables for Architects and Engineers in calculating the strains and capacity of structures in Iron and Wood. 12mo, morocco, tucks......... ..... 1 50

SCRIBNER (J. M.) Engineers' and Mechanics' Companion. Comprising United States Weights and Measures. Mensuration of Superfices, and Solids, Tables of Squares and Cubes, Square and Cube Roots, Circumference and Areas of Circles, the Mechanical Powers, Centres of Gravity, Gravitation of Bodies, Pendulums, Specific Gravity of Bodies, Strength, Weight, and Crush of Materials. Water-Wheels, Hydrostatics, Hydraulics, Statics, Centres of

Percussion and Gyration, Friction Heat, Tables of the Weight of Metals, Scantling, etc. Steam and the Steam-Engine. 16mo, full morocco..... ... .. 1 50

SCHELLEN (Dr. H.) Magneto-Electric and Dynamo-Electric Machines: their Construction and Practical Application to Electric Lighting, and the Transmission of Power. Translated from the third German edition by N. S. Keith and Percy Neymann, Ph.D. With very large additions and notes relating to American Machines, by N. S. Keith. Vol. 1, with 353 illustrations....... .... 5 00

SHIELDS (J. E.) Notes on Engineering Construction. Embracing Discussions of the Principles involved, and Descriptions of the Material employed in Tunnelling, Bridging, Canal and Road Building, etc. 12mo, cloth.......... .... 1 50

SHREVE (S. H.) A Treatise on the Strength of Bridges and Roofs. Comprising the determination of Algebraic formulas for strains in Horizontal, Inclined or Rafter, Triangular, Bowstring, Lenticular, and other Trusses, from fixed and moving loads, with practical applications, and examples, for the use of Students and Engineers. 87 woodcut illustrations. 8vo, cloth. 3 50

SHUNK (W. F.) The Field Engineer. A Handy Book of Practice in the Survey, Location, and Truck-work of Railroads, containing a large collection of Rules and Tables, original and selected, applicable to both the Standard and Narrow Gauge, and prepared with special reference to the wants of the young Engineer. Ninth edition. Revised and Enlarged. 12mo, morocco, tucks.... . .. . 2 50

SIMMS (F. W.) A Treatise on the Principles and Practice of Levelling. Showing its application to purposes of Railway Engineering, and the Construction of Roads, etc. Revised and corrected, with the addition of Mr. Laws' Practical Examples for setting out Railway Curves. Illustrated. 8vo, cloth.......... 2 50

—— Practical Tunnelling. Explaining in detail Setting-out of the Work, Shaft-sinking, Sub-excavating, Timbering, etc., with cost of work. 8vo, cloth.......... 7 50

SLATER (J. W.) Sewage Treatment, Purification, and Utilization. A Practical Manual for the Use of Corporations, Local Boards, Medical Officers of Health, Inspectors of Nuisances, Chemists, Manufacturers, Riparian Owners, Engineers, and Rate-payers. 12mo, cloth.......... .... 2 25

SMITH (ISAAC W.. C.E.) The Theory of Deflections and of Latitudes and Departures. With special applications to Curvilinear Surveys, for Alignments of Railway Tracks. Illustrated. 16mo, morocco, tucks. 3 00

—— (GUSTAVUS W.) Notes on Life Insurance. Theoretical and Practical. Third edition. Revised and enlarged. 8vo, cloth. 2 00

STAHL (A. W.) and WOODS (A. T.) Elementary Mechanism. A Text-book for Students of Mechanical Engineering. 12mo, cloth........................ 2 00

STALEY (CADY) and PIERSON (GEO. S.) The Separate System of Sewerage : its Theory and Construction. 8vo, cloth. With maps, plates, and numerous illustrations. 8vo, cloth............................ 3 00

STEVENSON (DAVID, F.R.S.N.) The Principles and Practice of Canal and River Engineering. Revised by his sons David Alan Stevenson, B.Sc., F.R.S.E., and Charles Alexander Stevenson, B.Sc., F.R.S.E., Civil Engineer. 8vo, cloth....... ....10 00

—— The Design and Construction of Harbors. A Treatise on Maritime Engineering. 8vo, cloth.......10 00

STILES (AMOS). Tables for Field Engineers. Designed for use in the field. Tables containing all the functions of a one degree curve, from which a corresponding one can be found for any required degree. Also, Tables of Natural Sines and Tangents. 12mo, morocco, tucks 2 00

STILLMAN (PAUL). Steam-Engine Indicator and the Improved Manometer Steam and Vacuum Gauges ; their Utility and Application. 12mo, flexible cloth... 1 00

STONEY (B. D.) The Theory of Stresses in Girders and Similar Structures. With observations on the application of Theory to Practice, and Tables of Strength, and other properties of Materials. 8vo, cloth........12 50

STUART (B.) How to become a Successful Engineer. Being Hints to Youths intending to adopt the Profession. Sixth edition. 12mo, boards.......   50

—— C. B.) C.E. Lives and Works of Civil and Military Engineers of America. With 10 steel-plate engravings. 8vo. cloth 5 00

SWEET (S. H.) Special Report on Coal, showing its Distribution, Classification, and Costs delivered over different routes to various points in the State of New York and the principal cities on the Atlantic Coast. With maps. 8vo, cloth 3 00

SWINTON (ALAN A. CAMPBELL). The Elementary Principle of Electric Lighting. Illustrated. 12mo, cloth..................................   60

SWINBURNE (J.) Practical Electrical Measurement. With 55 illustrations. 8vo. cloth.................... 1 75

TEMPLETON (WM.) The Practical Mechanic's Workshop Companion. Comprising a great variety of the most useful rules and formulæ in Mecnanical Science, with numerous tables of practical data and calculated results facilitating mechanical operations. Revised and enlarged by W. S. Hutton. 12mo, morocco...... 2 00

THOM (C.) and JONES (W. E.) Telegraphic Connections embracing recent methods in Quadruplex Telegraphy. Illustrated. 8vo, cloth............................. 1 50

THOMPSON (EDWARD P.) How to make Inventions; or, Inventing as a Science and an Art. A Practical Guide for Inventors. 8vo, paper.................... 1 00

TREVERT (E.) Electricity and its Recent Applications. A Practical Treatise for Students and Amateurs, with an Illustrated Dictionary of Electrical Terms and Phrases. Illustrated. 12mo, cloth.................. 2 00

TUCKER (Dr. J. H.) A Manual of Sugar Analysis, including the Applications in General of Analytical Methods to the Sugar Industry. With an Introduction on the Chemistry of Cane Sugar, Dextrose, Levulose, and Milk Sugar. 8vo, cloth, illustrated........ 3 50

TUMLIRZ (Dr. O.) Potential and its Application to the Explanation of Electric Phenomena, Popularly Treated. Translated from the German by D. Robertson. Ill. 12mo, cloth........ .... 1 25

TUNNER (P. A.) Treatise on Roll-Turning for the Manufacture of Iron. Translated and adapted by John B. Pearse, of the Pennsylvania Steel Works,

with numerous engravings, woodcuts. 8vo, cloth,
with folio atlas of plates...........................10 00

URQUHART (J. W.) Electric Light Fitting. Embody-
ing Practical Notes on Installation Management. A
Hand-book for Working Electrical Engineers—with
numerous illustrations. 12mo, cloth.................. 2 00

—— Electro-Plating. A Practical Hand-book on the
Deposition of Copper, Silver, Nickel, Gold, Brass,
Aluminium, Platinum, etc. 12mo.................... 2 00

—— Electro-Typing. A Practical Manual, forming a
New and Systematic Guide to the Reproduction and
Multiplication of Printing Surfaces, etc. 12mo....... 2 00

—— Dynamo Construction : a Practical Hand-book for
the Use of Engineer Constructors and Electricians in
Charge, embracing Framework Building, Field Mag-
net and Armature Winding and Grouping, Com-
pounding, etc., with Examples of Leading English,
American, and Continental Dynamos and Motors,
with numerous illustrations. 12mo, cloth............. 3 00

UNIVERSAL (The) TELEGRAPH CIPHER CODE.
Arranged for General Correspondence. 12mo, cloth.. 1 00

VAN NOSTRAND'S TABLE-BOOK. For Civil and
Mechanical Engineers. 18mo, half morocco......... 1 00

VAN WAGENEN (T. F.) Manual of Hydraulic Mining.
For the Use of the Practical Miner. 18mo, cloth..... 1 00

WALKER (W. H.) Screw Propulsion. Notes on Screw
Propulsion, its Rise and History. 8vo, cloth........ 75

WANKLYN (J. A.) A Practical Treatise on the Examination of Milk and its Derivatives, Cream, Butter,
and Cheese. 12mo, cloth.......................... 1 00

—— Water Analysis. A Practical Treatise on the Examination of Potable Water. Seventh edition. 12mo,
cloth....................................... 2 00

WARD (J. H.) Steam for the Million. A Popular
Treatise on Steam, and its application to the Useful
Arts, especially to Navigation. 8vo, cloth........... 1 00

WARING (GEO. E., Jr.) Sewerage and Land Drainage.
Large Quarto. Illustrated with wood-cuts in the text,
and full-page and folding plates. Cloth............. 6 00

WATT (ALEXANDER). Electro-Deposition. A Practical Treatise on the Electrolysis of Gold, Silver, Copper, Nickel, and other Metals, with Descriptions of Voltaic Batteries, Magneto and Dynamo-Electric Machines, Thermopiles, and of the Materials and Processes used in every Department of the Art, and several chapters on Electro-Metallurgy. With numerous illustrations. Third edition, revised and corrected. Crown 8vo, 568 pages 3 50

—— Electro-Metallurgy Practically Treated. 12mo, cloth 1 00

WEALE (JOHN). A Dictionary of Terms Used in Architecture. Building, Engineering, Mining, Metallurgy, Archæology, the Fine Arts, etc., with explanatory observations connected with applied Science and Art. 12mo, cloth.... .. . . 2 50

WEBB (HERBERT LAWS). A Practical Guide to the Testing of Insulated Wires and Cables. Illustrated. 12mo, cloth. 1 00

WEISBACH (JULIUS). A Manual of Theoretical Mechanics. Translated from the fourth augmented and improved German edition, with an Introduction to the Calculus by Eckley B. Coxe, A.M., Mining Engineer. 1100 pages, and 902 woodcut illustrations. 8vo, cloth................. . ...10 00
Sheep... ....11 00

WEYRAUCH (J. J.) Strength and Calculations of Dimensions of Iron and Steel Construction, with reference to the Latest Experiments. 12mo, cloth, plates.. 1 00

WHIPPLE (S., C.E.) An Elementary and Practical Treatise on Bridge Building. 8vo, cloth.... .... 4 00

WILLIAMSON (R. S.) On the Use of the Barometer on Surveys and Reconnoissances. Part I. Meteorology in its Connection with Hypsometry. Part II. Barometric Hypsometry. With Illustrative tables and engravings. 4to, cloth........... ... .15 00

—— Practical Tables in Meteorology and Hypsometry, in connection with the use of the Barometer. 4to, cloth.... .... 2 50

WRIGHT (T. W., Prof.) A Treatise on the Adjustment of Observations. With applications to Geodetic Work, and other Measures of Precision. 8vo, cloth... 4 00

—— A Text-book of Mechanics for Colleges and Technical Schools. 12mo, cloth............... .... 2 50

www.ingramcontent.com/pod-product-compliance
Lightning Source LLC
Chambersburg PA
CBHW021944190326
41519CB00009B/1131